THE CALIFORNIA CURRENT

THE
CALIFORNIA
CURRENT

A Pacific Ecosystem and Its
FLIERS, DIVERS, and SWIMMERS

STAN ULANSKI

The University of North Carolina Press *Chapel Hill*

This book was published with the assistance of the Wells Fargo Fund for Excellence of the University of North Carolina Press.

Library of Congress Cataloging-in-Publication Data
Names: Ulanski, Stan L., 1946– author.
Title: The California Current : a Pacific ecosystem and its fliers, divers, and swimmers / Stan Ulanski.
Description: Chapel Hill : The University of North Carolina Press, [2016] | Includes bibliographical references and index.
Identifiers: LCCN 2015038635 | ISBN 9781469628240 (cloth : alk. paper) | ISBN 9781469628257 (ebook)
Subjects: LCSH: Marine ecology—California Current. | California Current.
Classification: LCC QH95.3 .U53 2016 | DDC 577.709794—dc23
LC record available at http://lccn.loc.gov/2015038635

CONTENTS

ILLUSTRATIONS

FIGURES, MAPS, & TABLES

Figures

Maps

Tables

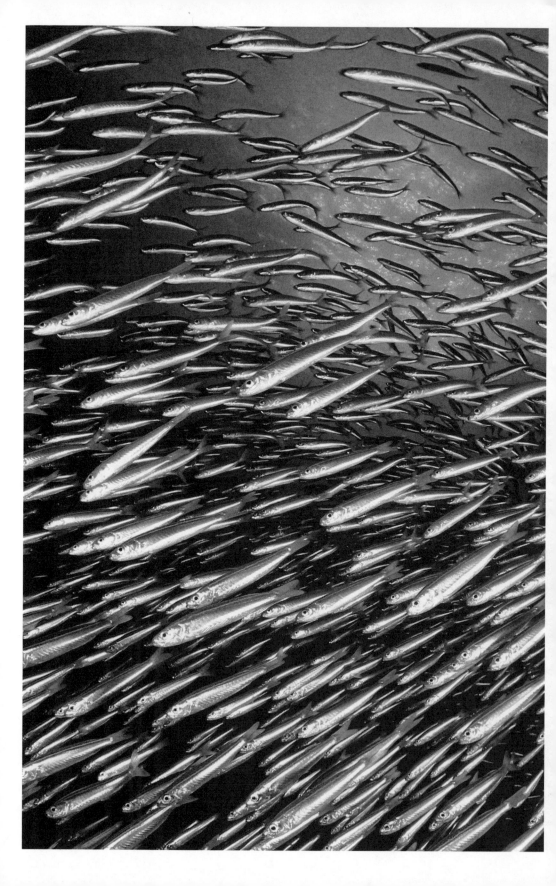

PREFACE

The California Current, part of the large, swirling North Pacific gyre, slowly flows southward along the Pacific coast of North America, stretching from southern British Columbia to the tip of Baja California, a distance of almost 2,000 miles. This languid current supports a vibrant marine ecosystem, from tiny plants to giant whales, from deep-water organisms often found miles below the surface to avian gliders that soar hundreds of feet above the waves, and from panicked prey to stealthy predators. Teeming with life, the California Current Ecosystem has been likened to the species-rich Serengeti Plain of Africa.

This dynamic system owes its high biological productivity primarily to seasonal coastal upwelling, which infuses the ocean surface with nutrients. In turn, these nutrient-enriched waters yield a rich broth of planktonic organisms that form the base of a complex food web. The California Current Ecosystem sustains active fisheries, both commercial and recreational, plays a vital role in the economy of myriad coastal communities, and is a magnet for life.

Every year whales, sharks, tunas, turtles, and seabirds migrate great distances, some thousands of miles, to feast upon the ecosystem's rich food resources. These nomads of the Pacific, as with the African wildebeests that are constantly on the move searching for fresh grasslands, may roam the California Current Ecosystem seeking out new, but ephemeral, meadows of life. Driven by the need to feed, apex predators, such as the California gray whale, salmon shark, and bluefin tuna, undertake extensive north-south migrations yearly within the ecosystem.

The California Current Ecosystem links together many strands from both the natural and cultural worlds. This book explores the science of this unique biological realm, examines its relationship to the wider Pacific, elucidates its many life-forms, both above and within the sea, and chronicles the development of industries that depended on it. My approach to the subject is intended to lead you on a journey of discovery, an odyssey that illuminates the complex and nuanced interactions of humans and the myriad organisms that inhabit this eastern Pacific system.

The elephant seal, for example, is a marvel of the biological world, capable of deep and prolonged dives into the abyss due to a specialized circulatory system that allows it to store large amounts of oxygen. But our relationship with these animals was also one of unbridled exploitation. During the nineteenth century, elephant seals were hunted almost to the brink of extinction for their blubber.

Though the Spanish explorers of the sixteenth century may have been the first Europeans to sail the California Current, as they searched for suitable harbors to support their far-reaching ventures, Charles Holder, a transplanted Easterner and keen observer of nature, was the first to recognize the fertility of the surrounding waters. Upon his visit in 1886 to the remote island of Catalina, he was awed by the abundance and diversity of marine life. To Holder, all the essential ingredients were in place to make this site a mecca for the pursuit of big, strong, and fast fish, such as tuna and billfish. One angler who made the pilgrimage to Catalina was Zane Grey. While probably better known for his tales of the Old West, Grey was relentless in his pursuit of swordfish, to the exclusion of all other angling opportunities.

As I will emphasize throughout this book, the California Current Ecosystem is not homogenous throughout its length; there are markedly distinct and different oases of life. It is made up of several overlapping smaller ecosystems—nearshore and offshore, pelagic and benthic—with widely ranging conditions from north to south.

Present-day anglers can most assuredly associate with Holder's vision and Grey's passion as they travel to faraway volcanic islands, including the Revillagigedo Archipelago in Mexican waters, to pursue giant yellowfin tuna, which routinely reach weights over 300 pounds. The complex interplay of current flow against these undersea mountains leads to a robust food chain that is supported by upwelled nutrient-enriched water.

Further to the north are the Farallon Islands, a cluster of granite outcrops located east of San Francisco. These islands are home to a diverse group of pinnipeds that seek refuge and breeding sites on the islands' rocky shores. Almost every year, a primal act plays out in the waters surrounding these islands: the arrival of great white sharks—powerful, efficient carnivores—to feed on these marine mammals.

In spite of the vitality and diversity of life that flourishes today in these waters, history tells us a stark story about the fragility of such resources. During the early part of the twentieth century, the sardine fishery off the Monterey coast seemed limitless, accounting for approximately 25 percent of the total seafood catch in the United States. But it was not to last. By the mid-1940s, the sardine population had plummeted mainly because of overfishing, the canneries that lined the waterfront closed, and a thick malaise settled over the community.

And the organisms within the California Current Ecosystem are still under attack, threatened not only by humans but also by a changing environment. Warming and acidification of the seas can lead to food web disruptions, shrinking habitat, and alteration of migration patterns of many species. That said, my book is not an environmental polemic but a dispassionate analysis of the problems besetting this ecosystem and the potential solutions, such as marine sanctuaries and marine protected areas, to mitigate these concerns. It is also not my intent to present an exhaustive guide to the myriad creatures of the California Current Ecosystem, an almost impossible task, but rather to concentrate on those species that have over time come to be viewed by scientists, naturalists, and laypersons as iconic species of the California Current. (I have included within the Bibliography a list of guide books, which I have found helpful, for those readers who want to pursue specific groupings or families of organisms in more detail.)

It is my hope for you, whether you're a blue-water sailor or an armchair adventurer, that this book reveals the many biological treasures of the California Current Ecosystem, their influence on the culture and history of the Pacific coast, and our human relationship to them, one filled with both dangerous and heroically good intentions.

THE CALIFORNIA CURRENT

Chapter One

THE BOUNTIFUL WATERS

In the late nineteenth century, Charles Holder, an accomplished author, sportsman, and naturalist, traveled from Massachusetts to California, drawn to the opportunities and adventure that a growing West offered. Holder initially settled in Los Angeles, but spurred on by an incurable curiosity of the natural world, he would visit the remote island of Catalina in 1886. As he roamed this rocky island, located about twenty-two miles southwest of Los Angeles, he was astounded by the diversity of marine life: seabirds whirling overhead, sea lions lounging on the wave-swept shore, sea bass and yellowtail frantically chasing bait, and whales breaching the sea surface. Nothing back in New England, where Holder had been quite content to cast a fly to a rising trout on the small streams that dotted the landscape, compared with this cornucopia of life. While most definitely not the first human to come in contact with such a great diversity of species—the island was originally inhabited by local tribes thousands of years before Holder's presence—Holder did recognize the link between the fertility of the nearshore waters and the abundance of life there.

Fishermen were also attracted to these waters and were so effective and indiscriminate in their fishing practices that Holder's little bit of Eden was no longer quite the paradise he first saw. Compelled to bring to light what Holder viewed as the unbridled exploitation of resources, he wrote that he was "amazed and horrified at the sight of men fishing with handlines from the beach, pulling yellowtail from twenty-five to thirty pounds as fast as they were able to cast." He argued to whoever would hear his words that this type of fishing was tantamount to wholesale slaughter, an inequitable match that pits man against fish. Little did Holder know that future generations would often turn a deaf ear to his pleas; the plunder would continue.

The Bigger Picture

Catalina Island is only a microcosm within a much larger world that was essentially foreign to Holder. Stretching nearly 2,000 miles from southern British Columbia to Baja California, Mexico, is the California Current Ecosystem—a dynamic, diverse, and biologically rich environment in the eastern North Pacific. Some have likened this marine ecosystem to the great African savanna ecosystem, which is home to over 600 different life forms. As in the vast grasslands of Africa, the California Current Ecosystem has a great array of flora and fauna, distinct oases of life, migration corridors, and specific feeding sites. From microscopic organisms to the world's largest creature, the blue whale, with a profusion of finned, feathered, and furred creatures in between, all find a home in this vibrant ecosystem.

A view from thousands of miles above Earth, however, would not yield any discernible boundaries. There is no obvious line that separates oceanic biomes as we often see in the terrestrial environment. A satellite image of Africa shows the marked transition from the Sahara Desert to the equatorial tropical rain forest—a shift from the bone white of the desert to the forest green. Changes within and along the California Current Ecosystem are more subtle and often require measurements of temperature, salinity, dissolved oxygen, and water flow, to name a few, to mark the current's boundaries. Sea surface temperatures, for example, vary along the coast, with Northern and Central California generally having cooler water (52°F to 55°F) than Southern California (63°F to 66°F). The Santa Barbara Channel area marks the "boundary" where these waters mix.

Within this ecosystem, the living organisms—plants, animals, and microbes—function in conjunction with the nonliving components of their environment. One vital abiotic component of this ecosystem is current flow. Ocean currents have often been likened to rivers on land, yet there are limits to this analogy. They differ from rivers on land in that their size and range dwarf even the mightiest continental rivers. Their "banks" are fluid ocean, not soil and rock. Their temperatures, cold or warm, tell of their origin. Their diverse habitats support a menagerie of plants, fish, marine mammals, sea turtles, and seabirds. To oceanographers, one of the overarching factors influencing species availability and movement is current flow. Currents are the sea highways that many species employ to reach their spawning and feeding grounds.

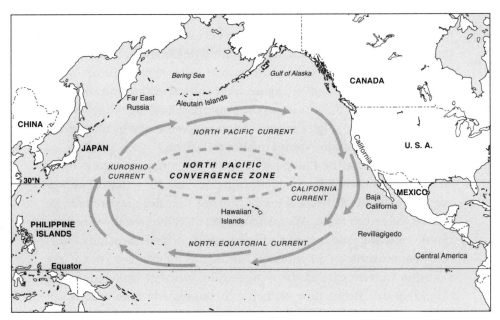

Ocean Currents in the North Pacific

The California Current is one of the world's four major eastern boundary currents. The current is part of the vast subtropical North Pacific gyre. This large, clockwise circulation, which encompasses an area of approximately 7 million square miles, is an extremely asymmetric disk of rotating fluid. Roughly the same amount of water that travels northward in the relatively narrow, powerful Kuroshio Current, the western limb of the gyre, is transported southward over most of the gyre. As the eastern edge of the gyre, the California Current languidly flows southward along the Pacific coast, transporting only a fraction of the total amount of water moving equatorward within the gyre. The forces that carry the California Current through the nets of fishermen, moving water hundreds of miles down the Pacific coast, were generated weeks, months, or even years earlier by winds blowing across the vast North Pacific Ocean.

Gyre Dynamics

Early explorers may have noticed that currents generally, but not always, flow in the direction of the wind. During the sixteenth century, Spain, a leading maritime power, commissioned an expedition to the

Philippines to cement its relationship with far-off trading outposts in the western Pacific. Commanded by Miguel López de Legazpi but under the guidance of the superb navigator Andrés de Urdaneta, the flotilla set sail from the port of La Navidad (near present-day Acapulco) on November 21, 1562. As proposed by Urdaneta, the ships took the most southern route, near 10° north latitude, as the outward-bound leg of the voyage. At this latitude, the ships were pushed steadily westward by the prevailing trade winds and the North Equatorial Current.

Upon arrival in the Philippine archipelago, Urdaneta had demonstrated, without a shadow of doubt from the crew, his wealth of knowledge of the immense tropical Pacific. But one herculean task remained: finding the return route. After departing the Philippines, the tiny fleet, again under the guidance of Urdaneta, sailed in a northeasterly direction to approximately 40° latitude. During this month-long voyage, the Kuroshio Current carried them northward. Whether Urdaneta recognized that this strong flow seemed to be unrelated to the wind is not clear. After being at sea for almost three months, the exhausted crew, suffering from hunger and other maladies, sighted the island of Santa Rosa off the California coast—the climax of the first Pacific crossing from west to east. (Some have claimed the Chinese may have been the first to make this transoceanic voyage.) Buoyed by this sighting, they sailed south along the coast, pushed along by the last link in the Pacific subtropical gyre—the California Current.

More than 200 years after the epic voyage by this hardy band of Spaniards, Captain James Cook of the British Royal Navy sailed into the western Pacific on a mission to find the fabled Northwest Passage. Cook, considered by historians to be a first-rate scientist as well as an explorer, located and meticulously documented the main branch of Kuroshio Current. But his logs are devoid of any explanation of what drives ocean currents.

Although detailed charts depicting the main patterns of ocean circulation became more prevalent during the nineteenth century, a rigorous scientific explanation of these unique gyres was missing. What caused these large flows? What was the relationship between air and water currents? Why were western boundary currents, such as the Kuroshio, fast and deep flowing, whereas eastern boundary currents, like the California Current, slow and shallow? Ideas and speculations were not limited to a determined band of ocean researchers but sprang from many segments of society, including cosmologists, philosophers, and the clergy.

Many theories were proposed, rigorously scrutinized, and ultimately abandoned. It would not be until the twentieth century, with more than a few wrong turns along the way, that oceanographers would come to illuminate the complexities of gyre dynamics.

The fundamental idea of wind-driven ocean currents would now be grounded on more quantitative analysis—a burgeoning area of study known as fluid dynamics—and scientific insight. Heretofore, ocean-ographers had only addressed through rudimentary observations and qualitative arguments questions concerning the effectiveness of winds in generating surface currents. But a simple observation would connect the fields of descriptive oceanography and fluid dynamics, ultimately yielding a unified and comprehensive explanation of ocean circulation.

During his epic voyage to the Arctic (1893–96), the Norwegian explorer Fridjof Nansen observed that ice floes drifted at an angle to the right of the wind, not in the direction of the wind. An answer to this puzzle was provided by the Swedish physicist V. Walfrid Ekman (1874–1954), who mathematically demonstrated that a steady wind causes a surface current to flow to the right of the prevailing wind in the Northern Hemisphere. The linchpin of his argument was the Coriolis force—a force that results from observing moving objects in a rotating system that accelerates them sideways from their original path (to right in the Northern Hemisphere and to the left in Southern Hemisphere). Ekman expanded upon his analysis by showing that the net transport of water (Ekman transport) within a water column is perpendicular to the prevailing wind flow.

Oceanographers now had two main components—Ekman transport and global winds—to develop their take on gyre formation. Leading the way was Harald Sverdrup, who in 1947 showed that the wind stress on the surface of the open ocean drives a north-south Ekman transport. The upshot of this argument is that in the North Pacific, as well in the other ocean basins, there is a net water movement approximately northward from the northeast trade winds, or perpendicular to the right of these winds, and approximately southward from the prevailing wester-lies in the higher latitudes. This convergence of water in the latitudes between these two wind bands drives a small increase in the sea surface height (approximately three to four feet).

And yet there is a limit to the elevation of this mound; the "hill" simply does not change into an ever-growing mountain. An equilibrium state is ultimately reached. The horizontal convergence of water near

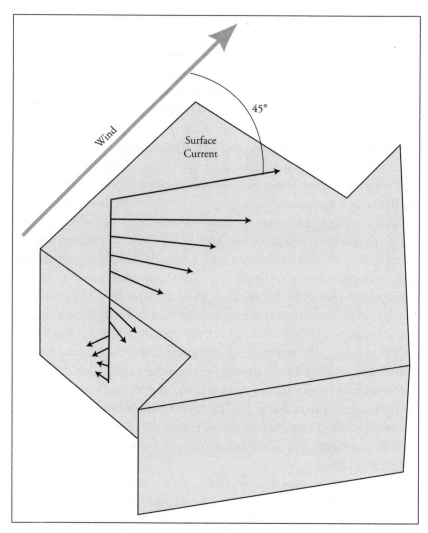

FIGURE 1 Ekman spiral and transport

the surface is balanced by the downward transport of water below the surface. But at this stage, you might ask, how does this relate to gyre formation? The pile of water in the center of the ocean has slopes and thus pressure gradients, which cause water to flow downslope, away from the mound. But because of the Coriolis force, the water that is flowing in response to the pressure gradient is deflected to the right. A balance of forces is attained where the flow of water is neither directed toward the hill nor away from it but instead *around* the hill. In this equilib-

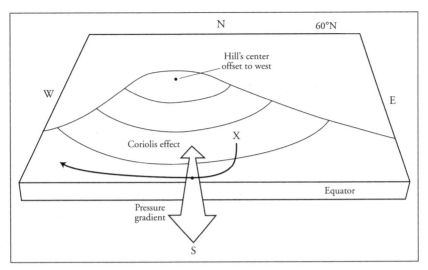

FIGURE 2 Hill of water and ocean gyre

rium state, the current moves in a circular, clockwise path around the mound. What conclusion can be drawn from this emphasis on forces and hills of water? The takeaway is that ocean currents within a gyre are not directly the result of wind drag on the ocean surface but result from the balance between the movement of fluid from high to low pressure and the deflection due to the Coriolis force.

Although the above discussion of ocean circulation elucidates the main features of gyre dynamics, it cannot account for the intensification of currents found along the western boundaries of the ocean basins. In an attempt to explain the western intensification of such currents as the Kuroshio, in 1948 Woods Hole oceanographer Henry Stommel developed a mathematical model of ocean circulation. Stommel found that the apex of the hill is displaced westward from its original location in the center of the ocean basin. Compressed against the western margin, the hill is distorted in appearance, resulting in a steeper slope on the western edge and a gentle gradient on the other side of the ocean. The main result is that water flows slowly south toward the equator across a wide swath of the eastern side of the ocean basin due to the gentle slope of the offset hill. In the Pacific Ocean, this flow is the California Current. In contrast, along the western edge of the basin, the water surges poleward rapidly in a narrow corridor in the form of fast currents, like the Kuroshio.

The California Current System

The California Current is part of a complex of ocean currents known as the California Current System, which includes the California Current, the Davidson Current, the California Countercurrent, and the California Undercurrent. This point is important because the California Current Ecosystem is really the ecosystem of the broad and diverse California Current System, not just the California Current.

The dynamics of the California Current System occur on multiple scales in space and time. For example, the aforementioned large-scale gyre and mean southward flow of the California Current are representative of long-term, average conditions, while smaller, transient features, such as eddies, disrupt and distort the mean flow. In practice, when one looks at the ocean surface from space, eddies are much easier to see than the mean southward flow. These eddies are like high- and low-pressure systems embedded in the atmospheric flow across the United States.

Due to its broad nature (360 miles in width), the California Current is viewed as having three distinct regimes: oceanic, coastal, and intervening transition zone. The oceanic region consists of the mean southward flow of the wind-driven subtropical gyre. The core of this flow, which is found well offshore, is characterized by low salinities and speeds. But within the coastal zone, the circulation can be quite complex, the dynamics somewhat different from that offshore. Here, poleward and equatorward flows occur that change over space and time. Transient eddies swirl clockwise or counterclockwise, like tops spinning on the ocean surface. At any given time, a cork caught in an eddy might easily travel north, south, or east as well as west. While high flow variability characterizes the coastal zone, the oceanic region is less active.

From Oregon south to Point Conception, California, the eastern edge of the California Current flows near the continental shelf break—the transition between the relatively shallow continental shelf and the deeper continental slope. South of Point Conception, the current flows approximately 100 miles offshore of Southern California. This indented stretch of the coast, similar to a giant bay, is known as the Southern California Bight. Upon flowing past Point Conception, the California Current brushes up against the relatively stationary Bight water, generating a large (80 miles across), counterclockwise swirling mass of water—the Southern California Eddy. The inshore component of this eddy flows between the coast and the Channel Islands and is commonly

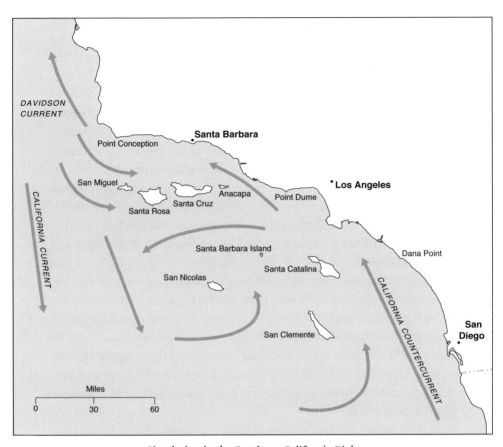

Circulation in the Southern California Bight

known as the California Countercurrent owing to its northward flow. During fall and winter, this poleward flow is present north of Point Conception, where it is called the Davidson Current.

The great diversity of life that Holder observed off of Catalina Island is due in part to the influx of nutrients carried southward by the California Current into the Southern California Bight. Microscopic organisms, which Holder, ever the inquisitor, came to see as the quintessential foundational organisms of a multitiered food web, flourish in this rich nutrient broth.

Grasses of the Sea

The microscopic organisms that most life in the California Current Ecosystem depends on are collectively known as phytoplankton or microalgae, so tiny that a cup of seawater may contain millions of these organisms. Phytoplankton are often mistakenly listed under the heading of plants but are plantlike in nature because they contain chlorophyll and have the ability to photosynthesize.

Phytoplankton are ancient, dating back more than 1.5 billion years. But over time, different species took their turn, kicking in under optimal conditions. By the Mesozoic Era (251 to 65 million years ago), three principal planktonic organisms (diatoms, dinoflagellates, and cocolithophores) would rise to prominence and dominate the modern seas.

Within the California Current System, diatoms are the main photosynthetic machines, converting the sunlight they absorb into usable chemical compounds. Their rate of photosynthesis depends on a number of environmental factors, including the amount of sunlight and the availability of nutrients, but is the highest among the planktonic species. While diatoms dominate the coastal zone of the California Current, dinoflagellates—endowed with small, whiplike appendages for limited vertical movement—may also be found there at some times of the year. Nanophytoplankton, minuscule unicellular organisms, can be found farther offshore. What they lack in size, they make up for by their high biological productivity. Together, these organisms form the base of a complex food web that ultimately supports organisms as big as the massive blue whale.

Phytoplankton do not live very long, from a few weeks to months, as compared with terrestrial plants that have life spans in years. The result is that the standing stock of plant biomass in the ocean is a thousand times less than on land, even though the global productivity of the ocean is comparable to that on land. Because the abundance of phytoplankton can change rapidly owing to a host of environmental variables, they tend to be quite variable both temporally and spatially. As if out of nowhere, meadows of life may appear, only to disappear over time.

The link between living organisms and their watery environment are the biogeochemical cycles occurring within the oceans. And none is probably more important for life than the nitrogen cycle in which elemental nitrogen, a critical component of plant chlorophyll, is converted into usable forms, such as nitrates, for uptake by organisms. The conversion, initiated by a host of different bacteria, occurs on the ocean floor and in the water column. The nitrates—nutrients or fertilizers—are initially entombed in the deeper regions of the oceans, trapped there by water's thermal stratification that prevents vertical mixing between the deep and surface layers. The uptake of these vital nutrients by phytoplankton residing in the upper, sunlit layers of the ocean is dependent upon the initiation of vertical motion throughout the water column. Under conditions of adequate light and nutrients, some waters of the California Current System may experience a phytoplankton bloom of enormous portions, easily visible from space.

The Next Link in the Food Chain

Zooplankton, or animal plankton, can be categorized according to size: from nanoplankton to megaplankton, such as large jellyfish. Copepods, macroplankton, are shrimplike organisms that are sometimes called the bugs of the sea because there are so many of them, over 10,000 species. These organisms play an extremely important role in the California Current Ecosystem by feeding voraciously on phytoplankton.

Zooplankton exhibit quite a unique feeding strategy. During the day, a multitude of sea creatures reside within the depths of the ocean. Like vampires, these animals tend to shy away from sunlight. Many of these creatures are invisible to the naked eye, but when viewed under a microscope, a startling array of organisms, with myriad textures and shapes, is on display. Included in this menagerie are larvae of fish and eels, relatively large crustaceans, such as krill, and copepods.

Each evening as the sun sets over the Pacific, these organisms slowly ascend from depths of 1,500 feet, or even deeper, to the surface, a vertical migration comparable with any in the animal kingdom. The migration is in response to the need to feed, to feast on the tiny phytoplankton residing in the surface waters. Feeding under the cover of darkness, these organisms generally avoid detection by bigger predators. Unmolested, they forage continuously throughout the night, but with the approach of daylight, they reverse course, sinking to spend another day in the darkness. These intrepid migrators move at a snaillike pace of just a few feet per minute, taking them hours to complete their journey.

Biologists have known about this migratory behavior since the 1800s because their sampling nets came back fuller at night than during the day. But the extent and true nature of the vertical migration was not discovered until World War II when the U.S. Navy was testing sonar equipment to detect enemy submarines. On many of these recordings, a puzzling sound-reflecting surface appeared that surfaced at night and returned to the depths during the day. Biologists later confirmed that the strong reflections were due to sound waves bouncing off of the bodies of countless marine animals. This densely packed layer of organisms is commonly referred to as the "deep scattering layer" that looks like a solid surface hanging in midwater.

Caught in the Middle

An integral, but often unappreciated, component of the California Current Ecosystem is a diverse group of fish that occupy the "middle" of the food chain. These species, such as the Pacific sardine and the northern anchovy, eat planktonic organisms and are prey items for sea birds, sea lions, whales, sharks, salmon, and tuna. Collectively, they are known as forage fish—small, pelagic fish that remain at the same level in the food web for their entire life cycle and are important food items for higher-level predators during their adult phase.

In addition to the Pacific sardine and the northern anchovy, other forage fish residing within the California Current Ecosystem include the Pacific herring, Pacific saury, lanternfish, Pacific sand lance, and smelt, along with numerous less-well-known species. These fish may occur throughout the ecosystem or in some cases have particular home ranges, such as the sand lance in Washington and the grunion in Southern California.

Their availability, or lack thereof, has been shown to affect the vitality and size of predator populations. Prey availability refers not only to food abundance but also timing, spatial distribution, and size classes, all of which may impact a predator's ability to locate and consume food. Salmon, for example, consume a variety of different forage fish, including anchovy, sardine, herring, and smelt, at different times of the year and at various stages of their life cycle. Small salmon smolts, for example, entering the ocean for the first time have prey size limitations. Not being able to find suitable prey during this critical time can severely impact their survival. Seasonal availability of forage fish may also be a key factor for other species. During the herring spawning season, Steller sea lions primarily concentrate their feeding efforts on these fish; studies have shown that at this time herring may make up 90 percent of their diet.

The California Current Ecosystem has historically undergone large fluctuations that impact forage fish abundance, often resulting in predator-prey mismatch when the timing or spatial distribution of forage availability differs from that of predator needs. Anchovies and sardines, for example, are known to ecologically replace each other as the environment changes. When one species has been plentiful, the other has usually been found at a reduced level of abundance, and vice versa. Sardines have tended to be abundant during periods of warmer sea surface temperatures, while anchovy numbers have been low. The abundance flips-flops when the ocean temperatures have cooled.

In contrast to other eastern oceanic regions, such as the Humboldt Current off of Peru that is dominated by a few or even just one forage species, the California Current Ecosystem has a high diversity of these organisms, although sardine and anchovy are the dominant species. A diverse forage assemblage can provide opportunities for prey switching, particularly when a favorite prey item may not be available. The high degree of forage diversity in the California Current Ecosystem precludes having a mid-food-web bottleneck or "wasp-waist" structure (high diversity of organisms at the bottom and top of the food web but low diversity in the middle) that is characteristic of the Humboldt Current. In this structure, predators do not have other food options if the timing or spatial distribution of the one or two food items does not match their needs.

A Cycle of Change

Over the decades, scientists have come to have a better grasp on how changes in physical conditions in the ocean affect marine life. In particular, oceanographic changes have a profound effect on the number of individuals of a species that make it through the larval stage to join the juvenile and adult population—a process known as recruitment. For many species of fish, there is a critical period that occurs between when the larval yolk sac (the main source of nourishment) is completely absorbed and when the small fish initially start to actively forage for food. At this time, the fish must have an adequate food supply as well as suitable habitat. If these conditions are absent, then the result is poor recruitment, a population that is in decline.

Along most of Central and Northern California, three periods occur during which winds and currents change seasonally: a spring/summer "upwelling period," a summer/fall "relaxation period," and a winter "Davidson Current period." The initiation of coastal upwelling is marked by an increase in northerly winds blowing generally parallel to the coastline. Wind stress on the water and the rotation of Earth result in surface water moving offshore (Ekman transport) that is replaced by cold, deep, nutrient-laden water. During the upwelling season, coastal waters lying over the relatively shallow continental shelf and upper slope are cold and generally high in nutrients. In contrast, offshore surface waters are relatively warm and nutrient poor. But over the decades, satellite images and data from moored instruments have revealed the inhomogeneous nature of the transition zone between the coastal and offshore waters. Long filaments of cold water extend from the coastal zone to more than 120 miles offshore. Some of these cold tongues are associated with strong, narrow seaward currents. This freshly upwelled water "squirts" directly offshore, undergoing little or no alongshore displacement.

Although it may appear that upwelling is a consistent and predictable process with subsurface water moving uniformly upward, the reality is quite different. Upwelling intensity, duration, and timing are highly variable within the California Current System, changing from year to year, depending on the strength and duration of the north to south winds. As recently as the summer of 2014, sea surface temperatures were 5°F to 6°F warmer than average—among the warmest temperatures recorded over the last thirty years. Winds that normally blew from the north decreased markedly and resulted in suppression of upwelling. Surface waters, in turn, heated up under the strong summer sun. As of now, the cause for the slackening of the usually reliable northerly winds remains a mystery.

The spatial variability of upwelling along the California coastline is affected by coastline geometry and bottom topography. Satellite images of sea surface temperatures reveal that the southern sides of promontories, including Cape Blanco in Oregon, California's Cape Mendocino, and Points Arena, Reyes, Año Nuevo, and Sur in California, exhibit strong upwelling events. In contrast, "upwelling shadows"—patches of warm water—tend to occur in the lee of coastal recesses (bays). The shadowed area reflects a nearshore region where upwelling is absent, and warm water is retained inshore.

Upwelling plays an important role for the rich marine resources of the California Current Ecosystem. The primary biological effect of upwelling is increased phytoplankton productivity, dominated primarily by only a few species of diatoms that rapidly reproduce because of high nutrient availability. The phytoplankton bloom supports large and diverse animal populations higher in the food chain, including fish, marine mammals, and sea birds. Although the eastern Pacific upwelling biome occupies only a minuscule part of the vast North Pacific Ocean, it contributes roughly 50 percent of the total fishing landings of this ocean.

Upwelling also plays a vital role in animal movement. The larvae of a number of species found within the California Current Ecosystem can potentially be moved a considerable distance offshore from their inshore habitats when surface water is moved offshore, as in cold filamentous currents, during upwelling events.

Depending upon the duration and strength of an upwelling episode, it can be a mixed blessing to the coastal environment. While upwelling can infuse the coastal waters with critical nutrients that fuel high levels of biological productivity, it can also deprive the nearshore environment of offspring required to replenish coastal populations, thereby impacting future fish stocks.

By late summer, the northerly winds diminish in strength or "relax," and upwelling events become more infrequent. At times, a total wind reversal may occur, resulting in onshore movement of clear ocean water into the coastal environs. Late fall marks the onset of the Davidson Current, which persists from December to February and accounts for an increase in coastal water temperatures. During many years, these periods may overlap extensively and do not recur with clockwork regularity, the timing reflecting changes in larger-scale atmospheric pressure systems.

The interactions between winds and currents influence temperature, nutrients, and the distribution of organisms and in the process create three distinct biogeographic regimes along the coast of California. The southern region, which includes the Southern California Bight, primarily supports warm-water fish and invertebrate species. Point Conception marks a transition point, where warmer Southern California water mixes with colder waters from the north. The second region, extending from Point Conception to Cape Mendocino in northern California, is characterized along the coast by seasonal upwelling events and offshore by the California Current. From Cape Mendocino north to, and at times beyond, the California-Oregon border is a regime of the coldest water and the organisms adapted to such an environment.

A Magnet for Life

The seasonal changes in water temperature and plankton abundance are the keys to the seasonal migration of many species to the California Current Ecosystem. Dr. Barbara Block of Stanford University has compared the diversity of marine life that migrates to this biologically rich environment to that of the great Serengeti wildebeest migration—the movement of vast numbers of wildebeest, accompanied by great herds of zebra and smaller numbers of gazelle, eland, and impala, in search of fertile foraging grounds.

Several species, including leatherback sea turtles, black-footed albatrosses, sooty shearwaters, salmon sharks, albacore and bluefin tuna, and elephant seals undertake annual migrations of hundreds to thousands of miles from the far reaches of the Pacific. The North Pacific Transition Zone is an essential habitat and sea pathway for a large number of these migratory species. This 5,500-mile corridor is one of the main oceanic features of the North Pacific and is a region of substantial temperature change that characterizes the transition between the warm waters of the subtropical gyre and the cold waters of the subpolar gyre to the north.

Biologists have noted a relationship between the migratory routes of many of the above marine megafauna with the position of the transition zone chlorophyll front—a sharp gradient in sea surface chlorophyll (a proxy for phytoplankton biomass)—in the North Pacific Transition Zone. High chlorophyll is found to the north of the front and lower chlorophyll to the south. The chlorophyll front shows a strong seasonal signal, shifting about 10° latitude throughout the year. It is at its southernmost position during the winter, around 30° north, but by summer it is at 40° north. During the winter, strong westerly winds push the subarctic nutrient-rich surface waters southward into the subtropical gyre, and the chlorophyll front lies along the southern edge of these waters.

Surveys have shown that while pre-spawning albacore tuna, two to five years old, are highly migratory and conduct transpacific migrations, spawning adults are confined to the subtropical and tropical zones of the central North Pacific. Their distribution is markedly influenced by the nature of the waters of the transition zone and its boundaries. During the spring and summer, adults capitalize on the productivity of the subarctic waters and the availability of nutritious prey. During winter, albacore retreat south of the subarctic front to warmer waters. In particular, catch data from the albacore fishery in the central and eastern North Pacific revealed that the highest catch rates were generally found along the chlorophyll front as the fishery moved eastward from the International Date Line to the North American coast in September.

Many of the highly migratory species that return to the California Current via the North Pacific Transition Zone initially return to the same foraging areas, showing remarkable site fidelity year after year. But for some, including yellowfin and bluefin tuna; mako, white, and salmon sharks; blue whales; and leatherback sea turtles, the journey is not done. These nomads of the Pacific may scour the California Current seeking out new, but ephemeral, oases of life. Driven not only by shifts in prey distribution but by oceanic process and species-specific thermal tolerances, these predators seasonally undertake extensive north-south migrations within the California Current. Other species move between nearshore and offshore waters, residing within the California Current before moving to distant points: the subtropical gyre and North Equatorial Current (blue and mako sharks and leatherback sea turtles) and the North Pacific Transition Zone (female elephant seals, salmon sharks, and Laysan albatrosses).

While many species are biologically programmed to migrate to and from the California Current Ecosystem, others are content to make it their permanent residence. California sea lions spend their whole lives in the California Current, using it for breeding, nursing, and feeding.

The findings showing that the California Current Ecosystem and the North Pacific Transition Zone attract and retain a diverse assemblage of marine vertebrates were made possible by the Tagging of Pacific Predators (TOPP) project. TOPP was a decade-long field component of the international program Census of Marine Life. Employing state-of-the-art electronic tagging devices, participants in TOPP deployed 4,306 tags on twenty-three species in the North Pacific, resulting in an unprecedented data set of 265,386 tracking days.

The collection of these data was more than a scientific exercise. Large marine organisms are apex predators in ocean ecosystems, and their depletion due to overexploitation by fisheries and climatic variability can have cascading effects on lower trophic-level organisms in both coastal and open-ocean waters, threatening marine biodiversity. The management and conservation of highly migratory species depends on deciphering how their movements relate to both physical and biological processes in the ocean; to improved understanding of ecological patterns in the context of these migrations; and to increased knowledge of the relationships between the organism and its environment, particularly if its habitat is changing relatively rapidly on account of natural and anthropogenic stresses.

A Variable Pacific

Numerous studies, ranging from days to decades in length, have shown a robust link between climatic variability and ecosystem change. These studies have focused on the natural frequencies of variability occurring in the eastern Pacific and the major components of marine ecosystems: biological production by photosynthetic organisms, forage species, and several trophic levels of predators.

At irregular intervals of three to eight years, for reasons still unknown, the normally persistent Pacific trade winds decrease in strength. This relaxation of winds is linked to changes in atmospheric pressure across the Pacific, known as the Southern Oscillation. But with these Pacific-wide alterations in wind patterns comes a dramatic response in the oceans. Warm water that has accumulated at the western side of the Pacific, the warmest water in all the oceans, now is free to move eastward, unrestrained by the normal trade wind pattern. The eastward-moving water, primarily centered about the Equator, reaches the coasts of Central and South America around Christmas. More than a century ago, Peruvian fishermen used the expression *Corriente del Niño* ("current of the Christ Child") to describe the anomalously warm water that had reached their shores. Over time, it was shortened to simply El Niño. What really caught the attention of the Peruvian fishermen was that this unusually warm water was associated with a virtual collapse of their productive fishery. The fish, in this case anchovies, just seemed to have vanished—a dire situation that often lasted for months. Unknown to these early Peruvians was that their catches plummeted owing to the depletion of nutrients in the surface waters when coastal upwelling was suppressed by the onset of an El Niño event, thus drastically decreasing biological productivity. The fish dependent upon the normally abundant food supply were now cast into the unenviable position of surviving, if at all, on a drastically reduced number of prey items or seeking out new, but distant, fertile waters.

The struggle between winds and currents in the equatorial Pacific can have ripple effects in the far reaches of the Pacific Ocean. During particularly strong El Niño events, warm tropical water floods north-ward along the Pacific coast, extending, at times, as far as Washington. The physical responses of the California Current System to an El Niño are multifaceted: stronger thermal stratification, as evidenced by a deepening and strengthening of the ocean's thermocline and weakening of coastal upwelling.

While some ecosystem responses to the physical changes in the environment are well known, including reduced nutrient input to the surface and decreased plankton biomass, others can be eye opening to both the scientific community and the general public. Where there are normally cold-water species, warm-water organisms now take hold.

During the late summer and fall of the 1997–98 El Niño, a particularly strong event, tropical, pelagic fish were found in the warm coastal waters off of Oregon, a region where they are rarely encountered. Mahi-mahi (dorado), yellowfin tuna, Pacific mackerel, and jack mackerel, to name a few, migrated from their tropical haunts in Mexico to Oregon, reflecting the extent of the warm-water intrusion. Albacore tuna, which normally reside 100 miles from the Oregon coast, feeding at the interface of the warm ocean water and cold coastal water, moved onshore and were being caught by anglers just a few miles from the coast. Reports of striped marlin and blue marlin catches from northern California to Oregon stunned the angling community. The blue marlin, in particular, is the most tropical of all the billfish, preferring surface temperatures between 74°F and 82°F, and is normally found below the California-Baja California border. But during El Niño, its geographic range increased more than 1,000 miles. In response to this climatic event, the blue marlin had undergone a distinct regime change, albeit for a limited period, until El Niño weakened and oceanic conditions returned to their normal state.

Probably the most unusual animal that invaded northern waters was the jumbo squid, which was initially caught as bycatch by Oregon-based trawlers targeting rockfish, which had also moved northward. This was the first time that squid had been reported in Oregon waters. Normally, this large, voracious squid is found in the eastern tropical Pacific, occasionally occurring off Southern California.

Squid were so abundant off Oregon from June through November of 1997 that commercial fishermen were targeting them for a lucrative market in Southern California. Landings in June alone were almost 35,000 pounds, in September 4,000 pounds, but by December were down to only 100 pounds. By early 1998, the migration of large pelagics had also ceased. El Niño had lost its grip on these northern waters.

While the influx of new species was a boon to the recreational and commercial fishing communities, there was a downside. Huge schools of ravenous mackerel roamed these northern waters in search of prey. And they found it in the form of juvenile salmon entering the ocean for the first time. Millions of salmon succumbed to the mackerel onslaught.

Displacement of endemic species was common. Cold-water species moved north in search of cooler water. Surface-oriented schooling fish—their thermal tolerance tested—sought relief deeper down in the water column. In all, almost two dozen species were reported in areas where they are not typically sighted.

At times, El Niño impacts can be complex, leading to conflicting outcomes. Sardines, for example, spawn in waters where the sea surface temperature is above 57°F. During the 1997–98 El Niño, their spawning habitat increased. But weak coastal upwelling led to less food for adults, contributing to poor egg production. In contrast, the following year, when upwelling was at full strength, was a banner year for egg production, despite the offshore displacement of warmer water.

Fish population disruption, both temporally and spatially, is only one effect of an El Niño. Its impact reverberates throughout the California Current food web, altering the vitality, abundance, and distribution of seabirds and marine mammals that depend on the forage items that have been displaced or depleted by the dearth of lower trophic organisms.

The Case of the Vanishing Fish

In the 1990s, Steven Hare, a fisheries scientist in the state of Washington, was scanning through old issues of fishing journals and was intrigued by a number of quotes:

Pacific Fisherman, 1915

"Never before have the Bristol Bay (Alaska) salmon packers returned to port after the season's operation so early."

"The spring (chinook salmon) fishing season on the Columbia River (Washington and Oregon) closed on noon at August 25, and proved to be one of the best for some years."

Pacific Fisherman, 1939

"The Bristol Bay (Alaska) Red (sockeye salmon) run was regarded as the greatest in history."

"The (May, June and July Chinook) catch is one of the lowest in the history of the Columbia (Washington and Oregon)."

National Fisherman, August/September 1972

"Bristol Bay (Alaska) salmon run a disaster."

"Gillnetters in the Lower Columbia (Washington and Oregon) received an unexpected bonus when the largest run of spring chinook since counting began in 1939 entered the river."

Pacific Fishing, 1995

"Alaska set a new record for its salmon harvest in 1994, breaking the record set the year before."

"Columbia (Washington and Oregon) spring chinook fishery shut down, west coast troll coho fishing banned."

Upon further digging into the subject, Hare found that for much of the past two decades before the onset of his study, Alaskan salmon fishers had prospered, whereas those in the Pacific Northwest had fallen on hard times. The only conclusion that Hare could arrive at was there exists a strong correlation between Alaska and Pacific Northwest fisheries. When Alaskan fishermen won, the fishermen from Northern California to Washington lost and vice versa. A question remained: what was the driving force behind this pattern of alternating salmon abundance? Hare and his colleagues had a hunch that these long-term fluctuations were the result of climate changes throughout the Pacific basin.

Pacific salmon, of which there are five species (Chinook, coho, chum, pink, and sockeye), are particularly sensitive to changes in their environment. Salmon are surface oriented and thus are metabolically constrained by surface ocean temperatures. With warming waters, salmon will seek out cooler waters as metabolic rates accelerate with warming. Were there heretofore undocumented long-term, basin-wide cooling and warming events?

In 1996, Hare coined the term "Pacific Decadal Oscillation (PDO)." The PDO is a long-lived El Niño–like pattern of Pacific climate variability. Although the two climate oscillations have similar spatial climate fingerprints, they have different behavior in time.

The occurrence of the PDO is marked by a warm or cold phase that may persist for decades. During the warm phase, sea surface temperatures in the northeastern and tropical Pacific are above average but below average in the western Pacific. During the cold phase, the reverse is true. Extended warm periods have seen enhanced coastal ocean productivity in Alaska, fueling salmon production, and diminished productivity off the coast of the Pacific Northwest. During the cold periods, the opposite north-south pattern of productivity is prevalent.

Unfortunately, causes for PDO are not currently known. Considerable controversy now exists over how it works and what mechanisms couple fluctuations in the PDO to changes in biological variables. These issues need to be resolved if we are to arrive at better models that accurately predict the impact of future PDOs on the Pacific ecosystem. The stakes are high.

Our Changing Planet

Robust, healthy ecosystems are always undergoing changes, fluctuating around a relatively steady condition. A sense of stability is achieved. However, with substantial "meddling," ecosystems may transition to a more variable state, which may be about the original equilibrium or about some new equilibrium.

The geologic record shows us that times exist when stability can be irreversibly damaged. Over the last 500 million years, five very distinct extinction events occurred, which, though markedly reshaping life on Earth, ushering in new, more adaptable players, were spaced out over tens of millions of years, allowing a certain degree of stability.

What lessons can be drawn from a critical study of the past? One answer is that cataclysmic spasms are an integral part of Earth's history and most likely will occur again. Some within the scientific community hold to the belief that we are already in a period of environmental upheaval, occurring at a pace not previously seen in the geologic record. As opposed to the five previous catastrophic events, the biosphere is now subject to massive forces initiated and enhanced by human activities. Earth may be reaching a time where ecosystems can no longer reach and maintain an equilibrium state, eventually leading to their collapse. Stability is replaced with future instability. Is this the stuff of Chicken Little?

A planet warming because of anthropogenic activities, particularly the release of greenhouse gases into the atmosphere at an unprecedented rate, may already be leaving its calling card in the marine realm. One study by Dean Roemmich and John McGowan of the Scripps Institution of Oceanography showed that over a forty-three-year span, the biomass of large zooplankton in waters off Southern California had decreased by 80 percent, while the surface layer warmed by more than three degrees, a disturbing finding not only because zooplankton are a significant link in the California Current Ecosystem's food web but also because these organisms are the main diet of some seabirds and many commercially important fish species. There are tantalizing hints that the plankton crash may have contributed to the 35 percent decrease in anchovy, sardine, mackerel, and squid landings since the 1950s. While the study raised eyebrows within the generally conservative scientific community and generated considerable controversy, the researchers caution that the question of causality has not been completely resolved. If the declines are a component of a natural cycle or cycles that may reverse in the future, then the long-term impacts on the marine biota may be minimal. But if the plummeting populations are a signal of anthropogenic forces at work or a natural trend of longer duration, then serious consequences are on the horizon. Either way, the California Current could offer an early peek at the changes that might be in store for other ocean regimes as warming continues its upward trend.

One take on the impacts of an overheated planet includes biota displacement, disruption, and death. But where would these impacts be the greatest? What organisms would suffer the most? Looking into a crystal ball to assess future scenarios is a risky proposition but a necessary one if officials are able to manage the potentially significant impacts on sea creatures. While not using a crystal ball, researchers affiliated with the Center for Ocean Solutions at Stanford University employed complex mathematical models and the long-term data set from the TOPP project to get a glimpse into the future. The picture was not a pretty one. If global temperatures continue to rise and biological productivity levels shift over the next century, critical ocean habitats, in particular the North Pacific Transition Zone, could move more than 600 miles from their current locations. This major displacement would have the greatest effect on some of the Pacific's top predators, including some sharks, elephant seals, and loggerhead turtles, with predicted losses of species diversity in this region as high as 20 percent owing to losses in critical habitat. For species that are already stressed, migrating longer distances, coupled with loss of foraging opportunities, would likely exacerbate population declines and slow or prevent any recovery. But one organism's loss is another one's gain. The outcome, as some have proposed, could be one of winners and losers. In the complex scenario of climate-related shifts, some animals, including seabirds and tuna, may reap benefits in the form of increased foraging habitat because of their wider thermal tolerances. Interestingly, the California Current System, a link to the North Pacific Transition Zone, would experience little or no adverse effect.

Andrew Bakun has argued that under the scenario of a warming planet seasonal upwelling and biological productivity would actually increase owing to the intensification of alongshore wind stress, driven by greater land-ocean temperature differences. In light of relatively recent climate simulations, Bakun's theory has come under fire. But the point is that we do not really know how the California Current System will respond to climate change.

Despite the decades of probing this environment, trying to unlock its secrets, our understanding is still evolving, an ongoing process that will require a combination of intensive field research and improved modeling efforts. The real danger may be reaching a tipping point that we do not recognize until it is upon us.

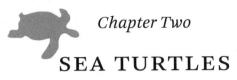

Chapter Two

SEA TURTLES
Living Dinosaurs of the Ocean

Far from the waters of the California Current, a massive female leatherback turtle ever so slowly shuffles up a remote beach in Indonesia. She has bred in the open ocean, but where exactly is not known. Her single-minded goal is to nest, to lay her eggs in the soft beach sands — a ritual that has been played out over the ages by her ancestors. The probability is high that the beach she has chosen is the same one from which she was hatched more than thirty years ago.

She finds a suitable site above the high tide line and diligently excavates a hole, big enough to hold the dozens of eggs she will deposit. The dark night sky offers her some privacy as well as shielding the nest from prying eyes. With the eggs carefully deposited, she sets about covering them up, using her big flippers like paddles to shovel the sand into the nest cavity and pack it firmly. Satisfied that her nest is now secure from predators, she returns to the sea. Ten days later, she will come back to the beach, excavate another nest, and deposit another clutch of eggs, a ritual she may repeat as many as eleven times during the nesting season. After she has laid her last clutch, her maternal responsibilities are done; the would-be hatchlings will have to fend for themselves. And unbeknownst to her, her offspring will enter a dangerous world of storms and predators. The large number of eggs laid may help ensure that some offspring survive, but Mother Nature will have the last word.

Her focus now is on a forthcoming long journey, thousands of miles across the Pacific. Though a powerful, deliberate swimmer, she will take months to complete her arduous migration. She will most likely ride the great subtropical North Pacific gyre to the California coast, where she will feed. But in two to three years she will again venture across the Pacific to the Indonesian nesting site, a cycle she will complete a number of times during her long life.

Leatherback turtle (amskad/Shutterstock.com)

Our nomadic turtle is one of five sea turtle species, including logger-heads, greens, hawksbills, and olive ridleys, which can be found within the California Current, particularly within the flow's southern reach. The leatherback is the only surviving member of the family Dermochelyidae; all other species belong to another group, the Cheloniidae. Sea turtles have been swimming the world's oceans for tens of millions of years, but recent genetic research suggests that leatherbacks may be the most ancient, dating back some 60 to 100 million years ago. Independent fossil records corroborate the above finding, showing that the leatherback lineage emerged contemporaneously with the dinosaurs.

The leatherback's earliest ancestor may have been *Archelon ischyros*, an immense sea turtle that was over fifteen feet in length and weighed over two tons. Seventy-five million years ago it swam the shallow, sunlit seas that covered central North America in pursuit of equally gigantic squid. This great turtle possessed features—stiff, winglike foreflippers and a smooth hydrodynamic form—which can be found in modern sea turtles. But *Archelon* and the leatherback share one common trait alone: that of relatively rapid growth. Estimates from the analysis of growth rings in the bones of dead leatherbacks show that the leatherback may increase in size a phenomenal 10,000 times in the span of a decade.

The leatherback is, on the average, the largest reptile alive. Though adult females routinely exceed 600 pounds, a male turtle, which had become entangled in a fishing net and ultimately died, weighed 2,015 pounds and had a flipper span of eight feet—a true giant mariner. Next in heft to the leatherback is the green turtle, reaching weights up to 500 pounds, followed by the loggerhead, hawksbill, and olive ridley, the smallest at about 100 pounds.

Ancient turtles, like many reptiles at that time, had teeth for grinding and chewing, but today's sea turtles, instead of teeth, sport horny beaks made of keratin—a hard, protein substance, similar to the protein found in fingernails. Evolutionary pressures have shaped the beak to reflect the dietary preference of specific species. Leatherback turtles have sharp points on both their upper and lower jaws for slicing into the gelatinous mass of a jellyfish. The green turtle employs its serrated beak for cutting and trimming the sea grasses that they are fond of. Loggerheads and olive ridleys have a large, thick beak for crushing and eating a variety of hard-shelled organisms. And the beak of the hawksbill, named for its narrow head and birdlike beak, is perfect for ripping into sponges, one of its favorite foods.

Though sea turtles are relatively large, most people, regardless of time spent on water, will never encounter them, the primary reason being sea turtles stay submerged for long periods of time diving and eating. During a twenty-four-hour period, they may stay under for a total of twenty-three hours. They are only visible when they stick their head out of the water to breathe.

Sea turtle occurrence is also spotty and irregular. Being subtropical to tropical in nature, they appear to be more abundant during those times when the ocean water is warmer than normal. Yet some are more prominent than others. The green sea turtle has been observed clambering onto land to warm itself by basking in the sun alongside seals and albatrosses. It is one of the few marine turtles known to exit the sea other than at nesting time. Leatherbacks, in contrast, are at times only sighted on whale-watching or seabird excursions. Leatherbacks live almost exclusively in open-ocean habitats, often far away from human contact. (The female leatherback that has migrated to the eastern Pacific will never set a flipper on California beaches.) Leatherback male turtles are even more reclusive. After taking to the water upon hatching, they will not return to shore again during their lifetime. Even the juveniles are elusive, rarely seen by humans until they reach almost adult size.

Decreased sightings may also reflect that there are simply not very many turtles swimming in the oceans. But it was not always this way. At one point in our history, the temperate and tropical oceans teemed with turtles; some estimates place the numbers in the billions. During Columbus's fourth voyage in 1503, he wandered upon two small islands and noted that the surrounding water was full of sea turtles. So thick were the turtles, he believed he could walk across the water on their backs. Vessels traveling to the New World that had lost their way could follow the noise of huge populations of sea turtles swimming along their migration route and find their way to the Cayman Islands.

But upon sighting a turtle, how does one identify the species? A sea turtle looks, well, like a turtle—head, shell, and limbs. Coloration of the shell can be variable. The green turtle (*Chelonia mydas*), for example, is not named for the color of its shell, which ranges from olive brown to black depending on its habitat, but for the greenish color of its skin. (Many casual observers and even some biologists view the East Pacific green turtle—due in part to its black coloration—as belonging to a subspecies of green turtles or to an entirely different species altogether.)

The key to solving this puzzle is the nature of the outer layer of the shell, or the carapace. Leatherbacks, for example, have a flexible carapace with six or seven ridges running the length of the carapace. For the other species, the body is covered with plates, or scutes. The number and arrangement of the scutes on the carapace enable differentiation among the sea turtles. Green turtles, for example, have four lateral scutes that do not overlap, whereas the slightly heart-shaped carapace of the olive ridley has six or more scutes.

The hawksbill (*Eretmocheyls imbricata*), considered by many to be most striking of all the turtles, owes its beauty, in part, to its intricately patterned shell. A hawksbill's shell color appears enhanced by the thickness of the translucent scutes covering its carapace. These tough scutes, which have a thickness of about a quarter of an inch, are impregnated with myriad colors: amber, cream, brown, black, and red. The colors form intricate patterns, ranging from radiating zigzags to overlapping bursts to irregular splotches. Given this variation, each hawksbill shell is probably unique.

Hawksbill turtle (Andrey Armyagov/Shutterstock.com)

The carapaces of some sea turtles are often host to a variety of plants and animals that attach themselves to the shell. Known as epibionts, these organisms include various types of algae, worms, sea urchins, and barnacles. Of the extant sea turtle species, loggerheads (*Caretta caretta*) may have greatest number and variety of organisms hitching a ride with them—well more than a 100 different species have been identified on the backs of turtles. Their carapaces are living mats of organisms, like an old log at sea encrusted with organisms. Most of these hitchhikers are harmless as they reside on the shell. Attached barnacles, for example, use their wispy legs to grasp at planktonic organisms flowing by. They do not feed parasitically off the turtle's resources, but their impact may not be completely neutral, particularly when they reside on the skin of the turtle. At this time, the scientific jury is still out with regard to the possible effect of these organisms on the health of the turtle. Taking into account the carpet of living organisms that partially cover the shell of the loggerhead and that its shell scutes are frequently fragmented and peeling, some might get the impression of the loggerhead as a weathered and worn turtle—a sharp contrast to the more strikingly beautiful hawksbill.

Loggerhead turtle (bikeriderlondon/Shutterstock.com)

Life Begins in the Sand

All sea turtles lay eggs, a trait they share with other turtles and most reptiles. As such, they are oviparous—producing eggs that develop and hatch outside the maternal body. Some reptiles retain their eggs in their bodies, and ultimately live offspring enter the world. But excavating nests and laying eggs must work for turtles, as the practice has been employed for millions of years.

Nesting is a solitary affair for most sea turtles: a single sea turtle hauls herself from the sea onto her chosen beach. But the olive ridley (*Lepidochelys olicacea*) is the exception, exhibiting one of the most remarkable nesting rituals in the natural world. Large numbers, in some cases, thousands, of olive ridleys gather offshore. Then all at once, as if spurred on by some primeval urge (what triggers this mass movement is still a mystery to the scientific community), they make a mad dash to the beach to nest. Like an amphibious landing, advancing waves of turtles crawl shoulder to shoulder, completely filling the lower beach. Those that pause to rest are crawled over or pushed aside by the more anxious ones.

The Spanish refer to these synchronized arrivals as *arribadas* (arrival by sea), literally a blanket of turtles laying thousands—or perhaps millions—of eggs. Playa Ostional and Playa Nancinte in Costa Rica are two of the world's major *arribada* sites. The annual nesting population at these two beaches has been estimated to be between 600,000 and 750,000 turtles. The largest *arribada*, which lasted for ten days, occurred in 1995 at Ostional, during which approximately 500,000 sea turtles emerged from the sea to lay their eggs. With a beach of about 2,500 feet in length, every foot of Ostional's shore was most likely occupied by scores of ridleys.

The eggs that are deposited in the sand are not like bird eggs, which have hard shells, but are pliable, similar to soft leather. The thin and porous shell facilitates the exchange of gases through it, supplying the developing embryo with oxygen and allowing carbon dioxide to escape to the atmosphere above.

All sea turtles nest on beaches, and how fast an embryo becomes a hatchling depends on the temperature of the beach sands. If the sand is cool, 77°F or so; the nest is in the shade; or there has been a prolonged period of rain, the incubation period is sixty-five to seventy days. But when the sands heat up to over 95°F, hatchlings may emerge from the eggs in only forty-five days. Because the female excavates a number of suitable sites during her nesting period, eggs laid on the same beach but in different places may hatch weeks apart.

The sand temperature also controls the sex of the hatchling. The offspring of green turtles, for example, are males if the temperature is 82°F, but when the temperature rises to 88°F, females emerge from the eggs. If the temperatures fall between these two values, a mixed clutch can result. (In a world that may be undergoing climatic perturbations not seen for eons, a warming environment can negatively impact sea turtle nests. Warmer temperatures result in more female hatchlings, while even higher temperatures may result in complete clutch failure. An ongoing study, initiated in 2005, has documented that nest temperatures are exceeding the tolerable limit for leatherback egg development, resulting in skewed sex ratios and low hatching success.)

Even the eggs themselves, in particular, the heat they produce and their orientation in the nest, may affect the sex of the hatchlings. Eggs in the center of the nest are warmer, most likely yielding females, whereas those on the periphery are cooler and more likely to produce males. The sex of the hatchling may simply come down to a random event—which

Green turtle nest (David Evision/Shutterstock.com)

way the egg bounces into the nest. To complicate matters, the middle of the incubation period is a critical time for sex determination. If the temperature drops a few degrees, embryos that were assuming female characteristics may end up as males—all due to the whims of nature.

In addition to the seventy to ninety eggs she will lay in her nest, a female leatherback deposits on top of the clutch many yolkless "eggs" that are globules of clear albumin packaged in a papery eggshell. The purpose of these infertile eggs has led to much speculation. Do they provide a moisture source for developing embryos? Do they cushion the eggs below as sand is piled on? Are they decoys for predators? Whatever the reason, the leatherback is an evolutionary success story.

As the embryos develop beneath the surface, danger is often not far away. Poachers, both human and nonhuman, may invade the nest; fungi can kill the eggs; and prolonged floodwaters may deprive the embryos of oxygen. Yet those that survive must face a whole new set of challenges.

Breaking out of the shell is the first order of business. And one turtle gets it all going. The tiny turtle, no more than a few inches in length, pierces its shell using an "egg tooth" on top of its beak. Others are stimulated by the commotion and commence to rip through their shells. And as James Spotila, who has observed and studied sea turtles for decades, notes, all hell then breaks loose, with the hatchlings crawling about, climbing over, and bumping into one another, in what he called "hatchling frenzy." In what has to be an exhaustive enterprise, the turtles, but not all, finally reach the surface after struggling for a day or even longer. Some are simply too weak to climb out of the egg nest.

The hatchlings' emergence coincides with nightfall. But it is not a time to rest; the sea beckons. The baby turtles must move quickly before they are discovered by a variety of nocturnal stalkers, including raccoons, cats, ghost crabs, and night herons. On *arribada* beaches, thousands of hatchlings swarm to the sea at once, often confusing their predators by their sheer numbers.

How the hatchlings immediately start moving in the right direction to the sea has long been a puzzle. In principle, their movement might be controlled by inherited information—genetic imprints—which instruct them to crawl toward a specific location. An interesting experiment performed a number of years ago tended to disprove this theory, showing that turtles instead used cues specific to their location to find the sea. The researchers relocated hatchling green turtles from the east coast of Costa Rica to the west coast, where the hatchlings crawled westward toward the sea, even though such a heading would have led inland at their original site. What environmental cues might lead turtles to the sea? Several options have received the most attention: more ambient light is reflected from the sea than the land, making that region brighter; the beach face slopes down in the direction of the sea, a clue to the turtle to head that way; and waves breaking on the shore might provide an auditory cue.

The intervening terrain between the nest and water can be intimidating to a one-inch hatchling. Flotsam, including driftwood and masses of seaweed, may seem like a wall, an impenetrable barrier. These obstacles must be navigated, generally by means of visual cues, if the hatchlings are to reach their destination. Even locomotion on sand can be tricky; loose sand can be unstable, resulting in slipping and impaired movement. While the turtles' limbs are supremely adapted for a life in the sea, their flippers also enable excellent mobility as they race over the sand. On loose sand, the hatchlings advance by using their flippers to push off a mound of sand that forms behind their flippers. But locomotion on sand is a balancing act for the turtles. High speeds, generated by large inertial forces, are needed to avoid the gauntlet of predators, but these same forces can result in failure through the fluidization of the sand. Moderation is the key—use enough force to push into the sand but remain below the threshold to break apart the ground. Near the water's edge on the hard-packed sand, the turtles employ a new strategy: digging a claw on their flippers into the sand to obtain purchase. The anchored claw allows them to thrust themselves forward. Their reward for all this effort? Backwash from a crashing breaker carries them seaward. They are now sea turtles.

Out at Sea

The new world that the hatchlings have entered can be quite intimidating—one of roiling surf and devoid of any visual markers for orientation. To maintain seaward heading early in their offshore migration, the hatchlings use wave propagation as an orientation cue. Since waves and swell generally move toward the shore and refract in shallow areas, swimming into the waves results in a sustained movement by the turtles toward the open ocean. In an hour or so of furious swimming, the hatch-

Hatchlings to the ocean (Kjertti Joergensen/Shutterstock.com)

lings may have traveled more than a half mile from the beach. Within the next day or so, the young turtles settle into a more reasonable swimming pace, content to be carried along by offshore ocean currents.

The sea will be their home for decades to come. But for years, biologists had very little information regarding the hatchlings' whereabouts, habits, or diet. In particular, the time between when they vanish in the surf only to reappear several years later is commonly referred to as the sea turtle's "lost years."

Archie Carr, viewed by many to be the "father" of sea turtle biology because of his groundbreaking field research on sea turtles and his unflagging enthusiasm for these creatures, pondered these turtle mysteries. When he died in 1987, more questions than answers were being raised by the cadre of biologists who followed in his footsteps. The journey of discovery would be a long and arduous one, but over time the mysterious sea turtle would reveal some of its secrets.

For loggerheads and leatherbacks hatched in the eastern Pacific, scientists have expressed some general opinions on the turtles' voyage through the Pacific Ocean. Their main conveyor will be the North Pacific gyre, carrying them from their natal sites thousands of miles across the open ocean. The turtles will be more than passive passengers swept along with the current, actively swimming for a few hours. In time, they are exhausted and must rest, opting to float in the current.

In this new offshore environment, safe from most but not all predators, the hatchlings will feed opportunistically, mainly on tiny planktonic organisms and larval shrimp and crab. A widely held belief was that the turtles would remain in the gyre currents and head toward California. Recent evidence, albeit indirect from a tracking study of loggerheads in the Atlantic, showed that some turtles frequently took detours, dropping out of the gyre altogether and seeking temporary refuge and respite in sea surface habitats, such as floating seaweed mats. The turtles were not simply traveling from one point straight to the next one but taking what they must perceive as necessary detours.

Upon reaching their preferred foraging grounds, the hatchlings will grow considerably in size. And with growth comes increased safety. The risk of death decreases markedly as maturing turtles graduate from a hatchling to juvenile to adult. By the time young sea turtles attain the size of a football, few predators could possibly make a meal out of them. Additional protection comes with a shell that hardens with growth, and some species, such as juvenile loggerheads, ridleys, and hawksbills, add another layer of defense, developing thick shell scutes with rear-pointing, spinelike projections.

For the juveniles, the transition from an open-ocean existence to making a living in their new shallow coastal habitats involves a significant change in lifestyle. They must shift from scavenging food anywhere near the surface to feeding on organisms that make their home on the bottom.

For some turtles, the change in diet is not easy. Young hawksbills, in particular, eat a mixed diet of invertebrates, including tubeworms, clams, and snails. But as adults, consuming sponges can be a tricky business. Eating a sponge's sharp, glasslike spicules and its store of chemicals would kill a human. The hawksbill needs time for its physiology and digestive tract to adjust to this new menu. Once it does, an average hawksbill adult, weighing about 150 pounds, will eat 1,200 pounds of sponges per year.

While both loggerheads and olive ridleys will feed upon a variety of shallow, bottom-dwelling invertebrates, ridleys can also be found offshore, spending a great deal of time chasing down pelagic organisms. Of all the turtle species, the leatherbacks are most reluctant to enter into coastal waters, content to forage in the open ocean on gelatinous organisms—their favorite prey items. Biologists have determined that leatherbacks reach adulthood without foregoing their pelagic lifestyle.

Regardless of their diet, sea turtles must control the saltwater intake that comes with the consumption of their food. When a hawksbill, for example, eats a sponge, the food initially resides in the esophagus, where the muscles around it contract to force the excess water that came in with the food out through the mouth. A sea turtle literally has to "burp" after eating. Another adaptation to prevent salt overdose are salt glands—large, modified tear glands near the eye. The opening of the tear ducts continuously excretes a thick, highly salty mucus, which gives the appearance of the turtle "crying."

The years will pass, perhaps a decade or more, before the turtles reach sexual maturity. For the female, it is time to find a suitor, mate, and head back to her natal beach.

Chief among these nomads are the loggerhead and leatherback sea turtles, logging more than 8,000 miles on their transpacific excursions from their feeding grounds to their nesting sites and in the journey defying exhaustion, storms, and even death. The sea turtle's mobility, large size, and hard shell are generally a deterrent to most predators, except for one top-level carnivore. Tiger sharks (*Galeocerdo cuvier*), a species of requiem sharks, are big, reaching lengths of over fifteen feet; stealthy, often launching their attack from below; and armed with a mouthful of sharp, serrated teeth, well suited for cutting through bone and shell. Although tiger sharks consume a variety of other prey items, including fish, marine mammals, and sea birds, they appear to have a taste for sea turtles. An analysis of the stomach contents of more than 400 tiger sharks showed that 21 percent of these sharks had large chunks of turtle in their stomachs. Are tiger sharks the only shark species that routinely feed on sea turtles? It appears so, because they often share the same habitat as the turtles and have developed a unique feeding behavior: after biting its victim, the tiger shark will characteristically slowly shake its head and body to enable the strongly serrated teeth to easily saw through the shell.

But why risk life and limb to reach that one special beach? It is a question that has intrigued researchers for years. Archie Carr was puzzled as to why green turtles that nest on Ascension Island in the south-central Atlantic utilize Brazilian feeding sites, more than 1,200 miles away. From the fertile minds of Carr and Patrick Coleman, a geophysicist, came forth a scenario involving sea floor spreading and natal homing. They proposed in 1974 that the ancestors of today's Ascension Island turtles had nested on islands adjacent to South America, not on those hundreds of miles away from the coast. The nesting occurred during the late Cretaceous Period, soon after the opening of equatorial Atlantic due to tectonic forces. The islands formed because of volcanic processes over the geologically active mid-Atlantic ridge. Over millions of years, these islands, riding on one of Earth's tectonic plates, were carried away from the ridge by sea floor spreading. They slowly eroded and sunk below the surface, to be replaced in time by new volcanic islands. According to the Carr-Coleman hypothesis, the ancestral sea turtles that could not find their "vanished" natal beach kept on swimming, ultimately reaching their new nesting site, another volcanic island. Eventually, even this island would submerge, and the turtles' journey would be extended ever farther—a process that culminates in their present-day migration to Ascension. The instinct to migrate to Ascension Island is therefore believed to have evolved over tens of millions of years of genetic isolation. The Carr-Coleman hypothesis may have generated more heat than light because while quite elegant in design—a meshing of biological and physical reasoning—there was no way at that time to test whether it was correct. The sea turtle enthusiasts would have to wait for the release of the genetic genie in the late 1980s. Mitochondrial DNA sequencing of green turtles from three rookeries showed extensive gene flow into the population, which was incompatible with the Carr-Coleman scenario of genetic isolation. In a nutshell, the genetic analysis showed that to a high degree of probability the colonization of Ascension Island has been evolutionarily recent.

The Pacific Ocean, particularly the central and western sectors, is dotted with hundreds of islands, some large and some barely more than specks of coral sand. In the tens of millions of years that sea turtles have roamed the vast Pacific, could it be that an island, which may have been an ancient nesting site, no longer exists and succumbed to the whims of a dynamic Earth? Would the turtles, as proposed by Carr and Coleman, doggedly continue on their odyssey for new nesting sites? These questions might be worthwhile exploring again in light of our new understanding of sea turtle biology and habits.

But to many sea turtle biologists, the answer to sea turtle migration lies not in the notion of disappearing islands but in ancient programming. A particular nesting choice reflects centuries-old conditions that made some sites preferable to sea turtles. Does the beach have the right slope so that the hatchlings can make their way to the sea? Is the texture of the sand loose enough to allow gas diffusion, yet dense enough to prevent collapse during digging? What predators, if any, lurk nearby? In the world of sea turtle real estate, the axiom "location, location, location" takes on a life-and-death meaning. In the end, turtle real estate decisions come down to historical experience, one acquired over long periods and passed down from generation to generation. The rationale is simple: a beach that worked for a mother turtle when she was a hatchling will work for her offspring.

Finding Their Way

A number of organisms, including salmon and sea turtles, exhibit a pattern of behavior known as natal homing in which they migrate away from their natal site and then return to reproduce in the same location where they began life. How sea turtles navigate in a vast, featureless ocean has come to light in recent years. For humans, latitude and longitude coordinates provide a reliable and accurate position fix on the earth's surface. But how do turtles determine if they are north or south (latitude) and east or west (longitude) of their desired heading? The answer is their ability to detect imperceptible gradients in the earth's magnetic field. The intensity of this magnetic field varies with latitude, strongest at the poles and weakest at the equator. If a turtle was thrown off course by a storm, it would be able to determine if it was north or south of its preferred heading by being able to detect and interpret the strength of the magnetic intensity and adjust accordingly.

For centuries, the determination of longitude proved to be more elusive for mariners than latitude, so elusive that it was long believed that it was impossible for animals to do it. While the intensity of the earth's magnetic field does not change traveling east or west, the pull's angle, or inclination, does change, even if only to an infinitesimally slight amount.

Although skeptics doubted that sea turtles use magnetic cues for navigation, at the "turtle laboratory," headed by Kenneth Lohmann, at the University of North Carolina in Chapel Hill, intricate experiments have shown that sea turtles have a compass for maintaining their direction that depends on their ability to detect subtle differences in both magnetic field intensity and inclination. To test the hypothesis that loggerhead and leatherback turtles can detect magnetic fields, Lohmann and his colleagues subjected juvenile turtles to different magnetic fields and monitored their response. Each hatchling was outfitted with a tiny nylon-Lycra harness and tethered to a rotatable arm in the center of a pool of water, which was surrounded by a large coil system that when turned on reversed the direction of the magnetic field around the turtles.

The turtles, unimpeded by the harness, could swim in any direction while a tracking system monitored and recorded the direction toward which the turtles swam. Half of the turtles were tested in the earth's magnetic field, and the other half swam under identical conditions except that the magnetic field around them was reversed. The results were encouraging. Most turtles swam steadily, maintaining a consistent course based on the magnetic field they were subjected to.

But while Earth's magnetic field may be sufficient to guide the turtles over basin-scale distances, a question remained: how did sea turtles find the exact beach on which they were hatched? A relatively new study by J. Roger Brothers and Kenneth Lohmann may have provided the answer. The turtles (loggerheads) imprint the unique geomagnetic signature of their natal site and employ this information to return years later. In other words, each part of the coastline has its own magnetic signature, which the turtles remember and later use as an internal compass. The turtle's commute to its birth site is not that straightforward because the magnetic field changes, albeit slightly, over time. The magnetic signatures drift slightly along the coastline. The aforementioned study was also able to show that subtle shifts in the magnetic field led to corresponding shifts in loggerhead nesting sites. Thus, if an adult female chooses her nesting site by seeking out the magnetic signature on which she was imprinted as a hatchling, she will invariably change her nesting location in accordance with the magnetic field changes along the coast. In addition, nesting turtles might routinely update their knowledge of the magnetic field of the nesting area each time they visit so as to minimize future navigational errors due to temporal magnetic variations. While it cannot be ruled out that nonmagnetic cues aid in navigation, not one, presently, has been identified that can provide the detailed positional information for long-distance navigation.

Sea turtles rely on other senses that serve them well in the more mundane tasks of finding food and avoiding danger. To succeed in these endeavors, turtles rely heavily on their vision. Judging by the anatomy of the eye, they see well underwater—a valuable trait considering the time they spend submerged—but are myopic in air. In particular, anatomical data show that compared with some blue-water fish, such as tuna and billfish, the eyes of sea turtles appear to be better adapted for bright light, which allows them to resolve detail in their surroundings.

Color discrimination also appears to be part of the turtle's visual arsenal but within a spectrum different from humans. All turtle species show color sensitivity toward the shorter wavelengths, such as violets, blues, and green, but the sensitivity decreases toward the red end of the spectrum. The ocean acts as a light filter—selectively absorbing colors with depths. The shortest wavelengths penetrate the deepest, not surprisingly coloring the blue world that turtles see for most of their lives. An exception may be the green turtle, which shows sensitivities well into the long wavelengths of light (red). This result may reflect its lifestyle—remaining mainly in the bright sunlit surface layers of the ocean—compared with other sea turtles.

Surprisingly, it turns out that some sea turtles, including greens, loggerheads, and leatherbacks, can also see ultraviolet light. This ability is spread across the vertebrate world, with some birds, other reptiles, and even some mammals able to detect light at ultraviolet wavelengths (shorter than visible light). It has been suggested that perceiving ultraviolet is a possible "secret communications channel" in sea turtles.

A casual observer might conclude that the absence of visible ears indicates that sea turtles lack the ability to hear. But a closer inspection of their anatomy reveals an auditory system that is both well suited to the ocean environment and quite sensitive. All reptiles, including sea turtles, have a single bone, which is found inside their head, in the middle ear that conducts vibrations to the inner ear. Turtles are most sensitive to low-frequency vibrations, which are essentially at the bottom of our own hearing range. While sea turtles do not appear to vocalize or use sound for communication, they may interpret the vibrations they receive to locate prey, to avoid predators, and to increase awareness of their environment. A pregnant female may determine that she is near her nesting site by picking up the low rumble of waves breaking against the beach.

Although little is known about a sea turtle's sense of taste, it is sensitive to touch, particularly on its flippers, and may use its tactile sense and acute sense of smell when foraging in murky water.

A Body in Motion

Sea turtles are excellent swimmers, well designed for long-distance travel. But seawater is a dense medium, a thousand times denser than air, which limits movement on account of drag. To reduce the influence of the water's drag, many marine organisms have a streamlined body.

For sea turtles, streamlining reaches its pinnacle in the design of the leatherback's shell. A broad anterior shell tapers, raindrop fashion, to a blunt point posteriorly, resulting in high hydrodynamic efficiency. To keep the speed-sapping effects of drag at bay, the fastidious green turtle may visit a reef "cleaning station," where its shell is kept free of epibionts by a number of reef inhabitants, including sheepshead bream, wrasse, and shrimp, happy to help out for a free meal, a mutualistic relationship that benefits all involved.

The ridges that run the length of the leatherback's carapace are believed to decrease turbulence, further enhancing the turtle's movement. Its body and shell are covered with a leathery but very smooth skin from which this species derives its name, *Dermochelys coriacea*, "the turtle covered in leather skin." Like the high-tech suits on an Olympic speed skater, the smooth skin reduces frictional drag from the fluid flow over and around the body.

The above combination of a streamlined shell, carapace ridges, and a relatively frictionless skin may make the leatherback sea turtle the most efficient of all the sea turtles at moving through the water. It is most assuredly the fastest. Studies have shown that swimming leatherback hatchlings have a lower drag coefficient than hatchlings of other species. To swim the same distance as other turtles of comparable size and weight, the leatherback expends, on the average, 20 percent less energy.

But for all sea turtles, efficient movement through a fluid involves trade-offs. Over time, turtles have given up some of the protections provided by their ancestral shell—an armor to ward off all but the most determined predators. Streamlining their bodies meant foregoing the shell space for their forelimbs and head. Open pockets beneath the top shell would create excessive drag, markedly decreasing swimming speed. In place of these empty spaces are smooth, powerful muscles that drive the propulsion strokes. To partially compensate for the vulnerability of exposed appendages to large-jawed predators, sea turtles have evolved robust skulls that are covered with dense bone, every bit as tough as the turtle's shell.

Though a sea turtle's morphology is a key component in efficient locomotion, where does the propulsion come from? Interest in propulsion in marine organisms dates back to ancient times. The first recorded accounts are from Aristotle, who believed a fish's pectoral and pelvic fins are used in a rowing motion to propel the fish through the water.

Although Aristotle's theory did not stand the test of time, his notion on the importance of anatomical structures in movement is still valid. In sea turtle locomotion, the focus is on the flippers and their use.

Such structures as flippers have a long evolutionary history; their basic design is similar to that of a foot. In sea turtles, the foot has evolved into a flipper that is sandwiched between two shells. But the flipper as a propulsion mechanism has over time become highly successful, found in a number of aquatic invertebrates. Though flipper design differs among aquatic species, it is similar within marine turtles. All sea turtles have semirigid foreflippers with flattened and elongated phalanges, the bones of the digits.

A sea turtle swims by simultaneously sweeping its flippers through the water in a figure eight pattern when viewed from the side. In the water, the flippers serve as wings and oars. During the forward movement—up and out—of the flippers, their orientation generates lift, similar to air flowing over a plane's wing, but since the lift works perpendicular to the orientation of the flipper, the turtle moves forward. As the flippers sweep back and down during their retraction, they provide forward thrust by pushing against the water like a paddle. For leatherbacks, the front flippers are more than half the length of their body, a span of six feet or more, and are powered by huge pectoral muscles, accounting for 30 percent of the turtle's body weight.

In order to turn, sea turtles change the amount of sweep of one front flipper and use the rudderlike movements of the hind flippers. For fast-swimming turtles, the rear flippers, which are just as broad and flat as the front flippers but only half the length, are constantly in motion, making small movements to adjust the course. Except for hatchling leatherbacks, which have very long foreflippers and swim very much like adults, most juvenile sea turtles use a dog-paddling technique, employing all four limbs simultaneously.

Sea turtles are such proficient swimmers that they rarely stop moving, slowing down only to bask on the surface or rest on the sea floor. This behavior has made them particularly unsuitable for long periods of captivity; they adjust poorly to the confines of a tank. This high mobility also makes them difficult to study at sea. But the deployment of electronic tags has opened a new world of satellite tracking of these nomads and monitoring their behavior.

Central to behavior of sea turtles is their ability to dive deep and stay submerged for long periods of time. The diving behavior is different among the species of sea turtles, reflecting both environmental and physiological adaptations. Loggerheads (and perhaps ridleys) spend a significant portion of their dive time on the bottom, possibly reflecting their foraging habits and choice of prey. They feed mainly on benthic-dwelling organisms, such as crabs and snails, using their large, thick beaks for crushing the hard shells of their prey. A loggerhead, in particular, is one tough customer. Mollusks and crustaceans are no match for this bruiser of the turtle world, given its huge head and large crushing jaws.

To the pelagic leatherback, exploitation of the water column is paramount, necessitating the ability to dive deeper than any other sea turtle, withstand crushing pressures, and adapt to different temperature regimes. With regard to diving, leatherbacks have no comparable peers in the marine reptile world. They are capable of reaching depths thousands of feet below the surface; in contrast, most other sea turtle dives are confined to the first few hundred feet of the water column. The leatherback's occasional sojourn into the abyss is not its normal diving behavior, which revolves around shallower dives but more time spent diving, as many as five dives per hour, day and night. The leatherback is almost frantic in its approach to diving: descend rapidly, rise quickly to the surface, grab a quick gulp of air, and immediately head straight back down again. What drives the leatherback to this incessant pattern of diving? It appears its behavior is in response to the availability of jellyfish, its favorite prey item. In tropical waters, jellyfish may be part of the deep scattering layer. The diving behavior of the leatherback is in sync with the diurnal movements of the layer: shallow dives during the night, when the jellyfish are close to surface, and deeper dives during the day, when these invertebrates sink from the sunlit layers. Its only respite from this frantic activity may occur at midday, when the deep scattering layer is at its deepest. Basking at the surface, the leatherback's black body absorbs the strong sun rays to moderate its body temperature.

The ocean depths to which a leatherback dives is a hostile environment—cold, dark, and one of crushing pressure. At 3,000 feet, water temperature has plummeted to a bone-chilling 40°F. The leatherback's large body size and thick, fatty insulation, and physiological adaptations allow it to control its body temperature in a manner that changes our whole view of sea turtles as "cold-blooded" reptiles. In fact, it has some traits that are more mammallike than reptilelike. A leatherback is more endothermic (the body can maintain a high core temperature) than poikilothermic (body temperature similar to that of its surroundings). A leatherback's core temperature may be sixty degrees higher than that of the surrounding seawater.

While immersed in this cold environment, the leatherback experiences almost 1,300 pounds of force on every square inch of its body, which could be deadly to a lesser physiologically adapted organism. Even at a depth of only about 100 feet, the plastron, the bottom shell, is pushed inward by the pressure at this depth. The leatherback, which has the most flexible and pliable plastron of all the sea turtles, is able to withstand the ever-increasing pressure with depth. Any internal air stores, such as in the chest cavity, are simply compressed without causing tissue or bone damage.

One of most insidious effects of deep diving is decompression sickness, more commonly known as the bends. During ascent, particularly during a rapid one, gases, mostly nitrogen, that passed into the bloodstream expand under the decreased water pressure and become lodged in the tissues, resulting in severe pain and, in some cases, death due to circulation blockage by the embolism. For years, decompression sickness was viewed as solely a human affliction, mainly occurring in divers breathing compressed air. But scientists also wondered if other organisms could experience decompression sickness.

Sea turtles manage gas exchange and decompression through anatomical, physiological, and behavioral adaptations. For example, sea turtles have relatively small lungs for their size, particularly true for the massive leatherback, and lungs that collapse under increased pressure. These traits drastically reduce the amount of gases that are exchanged, protecting the turtle from the agony of the bends.

Fossil evidence suggests that one ancient turtle, *Odontochelys semitestaca*, may have experienced decompression sickness. The bones of this Triassic (220 million years ago) marine reptile, which were analyzed by paleopathologists Bruce Rothschild and Virginia Naples, showed damage that is indicative of the bends. Both upper arm bones are pocked, a clue that blood flow around the bone was cut off because of the bends, effectively killing parts of the bone. While the researchers were hesitant to speculate as to how *Odontocheyls* received the bends, this unfortunate turtle most assuredly lacked the physiological adaptations that modern sea turtles possess to minimize this condition.

But new research shows that some modern-day sea turtles do, indeed, get the bends. A team of international scientists, led by Daniel Garcia-Parraga of Spain, diagnosed for the first time decompression sickness in a live air-breathing marine vertebrate—the loggerhead sea turtle.

The team concluded that sea turtles experienced decompression sickness after being brought to the water's surface too quickly. The turtles had become accidentally caught by or entangled in commercial fishing gear. Of the sixty-seven loggerheads studied, nearly half showed decompression symptoms, including unconsciousness and lack of mobility. In addition, gas embolism was observed postmortem in the tissues of eight turtles that had died.

The Case for Sea Turtles

Many within the conservation community support the view that all life possesses an intrinsic value—valuable in itself and not simply for its use—yet it can also be argued that sea turtles play a vital role in ocean ecosystems.

The active lifestyle of the leatherback requires food sources that provide the requisite energy and nutritional needs. How do leatherbacks meet these needs by consuming small, gelatinous jellyfish, which are 95 percent water? First, leatherbacks have been known to consume up to 440 pounds of jellyfish daily to extract the small amount of protein contained in each one. The leatherback is blessed with an extraordinarily long esophagus—extending from the mouth to the rear of the body before looping back to enter the stomach—which can expand to hold a great number of jellyfish at one time. The leatherback literally stuffs food down its esophagus.

As major consumers of jellyfish, leatherbacks play a pivotal role in controlling the jellyfish population. But the relationship has repercussions beyond these two species in terms of maintaining a balanced food web. As a result of overfishing in many areas of the world's oceans, jellyfish are gradually replacing once-abundant fish species. With declining fish stocks, jellyfish have less competition for food, resulting in an increase in jellyfish biomass. But the burgeoning population of jellyfish is proving detrimental to the recovery of fish stocks because jellyfish feed off fish eggs and larvae. Without leatherbacks and other turtle species that prey on jellyfish, a shift in species dominance is a real threat—one that may be irreversible if one species establishes a strong foothold within an ecosystem.

Green turtles are groomers, helping to maintain healthy seagrass beds that provide shelter and food for both recreationally and commercially important marine species. As herbivores, green turtles graze on seagrass blades. Without the constant cropping and recropping, seagrass beds may become overgrown with microorganisms, algae, fungi, and slime molds that may decrease the vitality and productivity of the beds. The green turtle's commitment to its landscaping chores also ensures it access to the nutritious new growth. It is not surprising, therefore, that green turtles may spend years within the same patch of vegetation, exhibiting strong site fidelity. Even when green turtles have been removed by researchers from their familiar surroundings, they invariably found their way back home.

Maintaining habitat also falls within the purview of hawksbill sea turtles. Armed with their beaklike mouths, they forage on a variety of marine sponges that find a home on coral reefs. Space on a reef is at a premium, and sponges compete aggressively for territory with reef-building corals. By consuming sponges, hawksbills control the sponge population, which if left unchecked would dominate reef communities, limiting the growth of corals and markedly changing composition of reef organisms.

Beaches, in general, are nutrient poor, lacking in nitrogen, phosphorus, and potassium. When females lay their eggs on these sandy shores, they introduce vital nutrients into the system, stimulating plant growth along the dune line. Increased vegetative growth not only helps stabilize the beach dunes but also provides food for a variety of plant-eating species. Even those eggs that are eaten by predators contribute to the nutrient input when these predators redistribute nutrients among the dunes through their feces. Sea turtles, by contributing nutrients to the beach ecosystem, are improving their own nesting habitat.

In addition to improving marine habitats, sea turtles provide habitat, particularly, as we have seen, to a host of epibionts. As sea turtles migrate from their feeding grounds to their nesting sites, they aid in the dispersal of some of these species, expanding the species's range and genetic diversity.

In the open ocean, miles from land, sea turtles become "floating hotels" for a number of seabirds and fish. Of all the sea turtle species, olive ridleys and seabirds have developed a unique relationship, particularly in the eastern tropical Pacific. When olive ridleys bask in the surface layers to warm their bodies, seabirds take advantage of this opportunity to perch on the exposed shell, possibly to rest or to seek refuge from attacks by sharks. Some seabirds will occasionally feast upon the epibionts on the turtle's shell. Also seeking protection, small baitfish may hover around the turtle in tight schools but often become prey to the ever-vigilant seabirds on the lookout for an easy meal. By providing a place to rest, feed, and hide, sea turtles are an important resource for other species within the Pacific ecosystem.

Humans and Turtles: An Uneasy Alliance

For thousands of years, sea turtles have played a role in human society and culture. Our earliest indication of the interaction of humans and sea turtles dates back about 7,000 years, when people inhabiting the Tigris-Euphrates Delta came upon sea turtles that had come ashore to nest. Soon the inhabitants of this ancient civilization were catching them in their fishing nets, and turtles, as verified by pieces of turtle bones excavated from numerous archaeological sites along the coast, became an important food source.

During the Bronze Age, about 6,000 to 4,000 years ago, sea turtles were viewed as a fine dining experience by the inhabitants along the Persian Gulf and Arabian Sea. A burgeoning cottage industry developed in which the coastal residents killed and butchered sea turtles, transporting the meat inland to the growing urban centers. It is very likely that royalty and other well-to-do residents feasted on green turtle steaks and maybe even consumed green turtle soup during holiday banquets. (Thousands of years later, even conservationist Archie Carr extolled the virtues of the rich, gelatinous, clear green turtle soup.) The popularity of sea turtles spread eastward across the vast Asian continent. Chinese texts dating back to the fifth century B.C. describe sea turtles as exotic delicacies.

As the years passed, turtles became cultural icons, often having spiritual and religious significance. In Hindu mythology, for example, the world rests on the sturdy backs of four elephants, which in turn stand on the back of a giant sea turtle residing in a vast ocean. When the elephants walked in a circle, Earth rotated during the day; when the elephants stumbled, earthquakes occurred; and when the turtle moved its giant flippers, it caused monsoons.

During the time of Babylonian dominance in the Middle East some 3,000 years ago, sea turtles adorned the walls of palaces, and about 2,700 years ago, Greek coins bore images of sea turtles. Greek legend has it that a shipwrecked sailor was carried to shore by hitching a ride on the back of sea turtle, nearly as big as he was. Greek chroniclers also report of Red Sea inhabitants catching huge sea turtles, opening their shells, and cooking the flesh with the sun's heat. No mention is made of the nature of the turtle and if it was edible. (From his trek in the waters around Baja California, John Steinbeck recounts in *The Log from the Sea of Cortez* that one of the crew had harpooned a hawksbill turtle, which they would ultimately consume: "The cooking was a failure. We boiled the meat, and later threw out the evil-smelling mess. Subsequently we discovered that one has to know how to cook a turtle.")

By 2,000 years ago, merchants developed markets, which stretched from ports in the Mediterranean south to the Horn of Africa and east to Indonesia, for trading turtle shells. But not all turtle shells are created equal. To many, the hawksbill is the most striking of all turtles, endowed, as we have seen, with richly patterned scutes that cover its shell. To the Japanese, the tortoiseshell or *bekko* became a much treasured commodity, to be used as decorative inlays in furniture and to make a variety of ornamental objects.

While the exploitation of sea turtles continued unabated for centuries on the Asian continent, turtles in the New World were also not spared, first by the indigenous populations then by the conquering Europeans. For many Native American societies, sea turtles were important components of their diet and culture. Archaeologists have determined that at least twenty-three sites in the United States, dating back 5,000 to 2,500 years, contained the remains of a number of turtles, including green turtles, hawksbills, and loggerheads.

In Central America, the Mayans not only consumed and traded in turtle meat but venerated turtles in their ceramics, stone altars, and figurines. The local artisans often fashioned gold jewelry into the likeness of turtles. At many locations throughout Costa Rica's southern

Pacific region reside more than 300 stone spheres. Some are small, the size of a bowling ball, and some are huge, fifteen tons in weight. No one has been able to determine exactly when they were made, who made them, and what purpose they serve. Explanations for the latter run the gamut from some sort of galactic map to a way of communicating with aliens. One novel explanation put forth by herpetologist James Spotila is that the spheres represent sea turtle eggs, which for centuries were a steady food supply for the indigenous population. Local craftsmen may have been commissioned to produce lasting symbols that pay homage to the great fertility of sea turtles that came upon the beaches. Whatever the reason for the existence of these spheres, one thing is clear: ancient societies for centuries viewed turtles as both a commodity and as cultural icons. Most of the indigenous populations had relatively little impact on the abundance of sea turtles, taking only what they needed for food and other uses. But burgeoning overseas markets in turtle products would severely deplete turtle numbers.

A Species on the Brink

About a hundred years ago, the fate of sea turtles began to change markedly. As a valuable source of protein, humans wantonly and, at times, illegally collected their eggs and killed turtles for their meat. Historically, the coastal communities of Mexico and Central America had the greatest impact on Pacific sea turtles. Villagers combed the beaches at night looking for a nesting female. These *velvadors* (those who stayed up at night) would immediately pounce upon an unsuspecting green or hawksbill turtle, rendering her helpless by flipping her over on her back. The next morning, the turtle would either be transported inland or carried out to sea to a waiting boat. Either way, the turtle was on its way to being killed.

Arribada events were particularly lethal for turtles because of the large concentration of nesting females. Sea turtles, as compared with their terrestrial cousins, cannot retract their heads and necks into their shells when they feel threatened. As a result, hundreds to thousands could easily be bludgeoned to death during the night.

Some turtle hunters might wait for the female to deposit her eggs before killing her. In a few days, the eggs that were gathered would be found dockside, waiting to be shipped to New York, London, and Tampa.

Improved transportation in the twentieth century meant that turtle products would arrive fresh at the markets, where they would be purchased by restaurateurs and wealthy patrons. The demand for turtle eggs and meat only fueled the unbridled poaching of eggs and killing of turtles.

Even today in Costa Rico, turtle eggs are prized as an aphrodisiac. So valuable are the eggs that poaching levels reach nearly 100 percent, with all the eggs removed from the nesting sites on the beach. Often, the sea turtle meat becomes an integral part of the observance of a religious holiday. In Mexico, for example, hordes of inland residents travel to the Pacific coast during the week preceding Easter in a relentless quest for sea turtles. During this brief period, as many as 5,000 turtles are killed.

The numbers of turtles harvested can be deceiving, giving the impression of a healthy, sustainable population year after year. Keep in mind one essential trait about sea turtles: their time scale. Most sea turtles have a generational span of many decades. During that lengthy period, every female turtle faces the possibility of being slaughtered on her nesting beach, without any evidence that the population is in a death spiral. Over the decades, maturing turtles enter the pipeline to adulthood and, like their predecessors, arrive at the nesting beach, giving the illusion of a boundless resource. The herpetologist Blair Witherington probably best sums up this dilemma: "Just as we might continue to witness the flicker of distant stars extinguished long ago, so too can we continue to take our fill of nesting sea turtles . . . for a time."

While the harvest of turtles for human consumption continues to take its toll on turtle populations, these marine reptiles face a host of new challenges that threaten their very survival. So great are the threats that presently all Pacific species are listed as endangered (a species is in danger of extinction throughout all or a significant portion of its range) or threatened (a species is likely to become endangered over the foreseeable future) under the Endangered Species Act.

The life of a sea turtle is, indeed, a perilous one, fraught with both natural and human threats. The Sea Turtle Conservancy has estimated that only 1 in 1,000 to 10,000 hatchlings will survive to adulthood. Several sea turtle populations are at an all-time low. The number of nesting leatherback turtles, in particular, has declined almost 6 percent per year since 1984, from the high-water mark of 14,455 nests in 1984 to 1,532 nests in 2011.

Though the leatherback was designated in 2012 as California's official marine reptile, a recent study supported by the Ecological Society of America arrived at a sobering conclusion: the species could become extinct in twenty years. Without a doubt, the primary threat to adult leatherbacks is from commercial fisheries. Because leatherbacks travel long distances across the Pacific, they run the risk of being caught in fishing gear, particularly longlines, thus undermining conservation efforts to protect them on their nesting beaches. At its core, longlining involves hanging a thousand baited hooks from a thick monofilament mainline that stretches tens of miles behind the boat. After a predetermined "soak" time for the bait, the gear is hauled back to retrieve the catch. While the targeted species are tuna and swordfish, the hooks do not discriminate as to what is caught. Global estimates of the impacts on leatherbacks of longlining operations are staggering—thousands are injured or killed each year.

George Shillenger of the Leatherback Trust organization and a host of researchers from around the world have identified a number of high-use sites in the Pacific—potential danger locales—for the leatherback. The data from turtles that were tagged and tracked from their nests in Indonesia showed that they traveled to a number of different sites, including those in the South China Sea, Indonesian waters, southeastern Australia, and the west coast of the United States. By fanning out, the turtles increased their chances to find adequate food stocks. But with the potential reward comes a risk: the turtles are more vulnerable to being caught by fishing gear in coastal and open-ocean waters. Knowing what sites turtles visit during their foraging outings is critical to limiting fishing activities in these areas at particular times of the year.

Analysis of tracking data of 186 loggerhead turtles, for example, yielded a strong association between the position of the chlorophyll front and the distribution of these turtles. This insight suggested an approach to reduce sea turtle–fishery interactions. In 2007, a project was undertaken to reduce loggerhead mortalities from longliners targeting swordfish and produced a near real time mapping tool, termed Turtle Watch. In a nutshell, the product provided longliners with a daily map of the position of the chlorophyll front and an area around it that represented the zone with the highest probability of interacting with a significant gathering of loggerheads.

But while there are some success stories, the task of implementing and enforcing conservation measures can be daunting. Far from the watchful eyes of resource managers, the vast, generally unregulated high-seas regime is, and has been, ripe for overexploitation. The highly lucrative international fisheries are generally reluctant to embrace any methods to minimize sea turtle mortality that also decrease fish catch rates.

Those leatherbacks that have migrated across the Pacific and avoided capture by longlines arrive along the California coast during the summer and fall to feast on the abundance of jellyfish. But another obstacle stands in their way: gill nets—large, rectangular mesh nets that ensnare fish by their gills but also trap other marine organisms. Stretching more than a mile in length and covering over 1 million square feet—the equivalent of twenty-one football fields—gill nets have become known as invisible "curtains of death." Recognizing the danger posed to marine wildlife by these nets, Washington and Oregon have not licensed fishermen to use drift gill nets since 1989. Californians in 1990 voted to eliminate the use of gill nets in state waters (from the shoreline outward to three miles) and restricted their use to deep water (defined as 6,500 feet). In 2001, federal protection for leatherbacks from the gill net fleet came with the establishment of the Pacific Leatherback Conservation Area, which stretches from Big Sur, California, to Lincoln, Oregon, and out 200 miles from the coastline. In this region, gill net fishing is prohibited from August 15 to November 15. As of this writing, the California state legislature is considering a total ban on gill nets.

But scientists are not simply waiting for the legal process to play out, which can be a long and arduous one, but are also looking at various strategies to decrease turtle bycatch in gill nets. One innovative approach is to take advantage of the sea turtle's ability to detect ultraviolet light. In an experiment off the coast of Baja California, Mexico, gill nets were illuminated with ultraviolet (UV) light-emitting diodes, with the hope that the UV-sensitive turtles would see the nets and avoid them. To the researchers' delight, the sea turtle capture rate was reduced by almost 40 percent in UV-illuminated nets compared with nets without illumination. But to the fishermen who participated in the project, one significant question remained: would the illuminated net reduce their catch? Another experiment was set up based on the fact that many commercially important fish are blind in the UV portion of the light spectrum. No differences were found in overall target fish catch rate or value between the illuminated and nonilluminated nets. While the initial results are promising, the efficacy of net illumination may be influenced by factors that are specific to other fisheries, including time of net setting, water transparency, and visual capabilities of other fish species. Nonetheless, collaborations must be established with the fishing communities in a manner that balances sea turtle conservation while maintaining a robust fishing enterprise.

The trading and selling of sea turtle products—skin (olive ridleys and green turtles), shell (hawksbills), eggs (leatherbacks, green turtles, and olive ridleys), and meat (green turtles)—has been a centuries-long tradition, and one that continues today owing to lax enforcement and a demand for these items. While clandestine, illegal trading activities undercut conservation efforts, the turtle trade is only a fraction of what it was before sea turtles were included in the Convention of International Trade in Endangered Species of Wild Fauna and Flora (CITES) in 1975. Under CITES all trade in sea turtles is strictly prohibited. Yet, despite this modicum of protection, CITES has as an Achilles' heel: it is a voluntary agreement that does not take the place of a country's national laws. Over the years, a number of countries have proposed changes to turtle protection under CITES. While most were rejected by other members of CITES or withdrawn, the pull of a lucrative turtle market may pressure some countries to opt out of CITES or, at the minimum, look for ways to circumvent it.

When Columbus set foot on the beaches of the Bahamas, they were as they had been for centuries before: pristine, relatively uninhabited, and clothed in darkness after sunset. Today, we find few beaches worldwide that have not been altered to some degree by the workings of humans. Coastal development results in artificial lighting that discourages females from nesting, forcing them to choose less optimal sites. Even those that nest on these light-polluted beaches may give birth to hatchlings that become disoriented upon leaving the nest, wandering inland toward the light instead of to their new home in the sea. Coastal armoring, in the form of seawalls, rock revetments, and sandbags used to protect prime beachfront property, alters the natural slope and composition of the beach—eliminating nesting habitat.

Without a doubt, the dangers to sea turtle survival are many, from climatic changes to anthropogenic impacts to natural threats. But with a little help and understanding from humankind, hopefully these living dinosaurs will be around for eons to come.

Chapter Three

SEABIRDS
Fliers, Gliders, and Divers

On September 14, 1492, still almost a month from his landfall in the Bahamas, Christopher Columbus noted in his log that the crew of the Niña had observed a tern, which Columbus interpreted as a good sign since this bird never wanders more than 25 leagues (75 nautical miles) from land. But the sight of land would prove elusive. While there are many species of terns—most of which prefer a coastal environment—only three or four species would likely have been seen this far out in the ocean. (According to Columbus's calculation, on this date he had sailed approximately 720 miles from his departure point in the Canary Islands.) One likely candidate may have been the Arctic tern, which during the late summer migrates from its breeding grounds near Greenland to the far reaches of the Southern Hemisphere, one of the longest migrations on record.

By late September, Columbus's crew had become despondent, feeling desperate about their plight. Perhaps to soothe their fears, on September 30 Columbus wrote, "Four tropicbirds came to the ship, a clear sign of land, for so many birds of one sort together show that they are not straying about, having lost themselves." Though the spirits of the crew were temporarily buoyed by this news, Columbus's reasoning would ultimately prove to be wrong—a vast, empty sea still remained in front of the three small ships. Ornithologists now know that only two species of tropicbirds are to be found in the latitude that Columbus was sailing, and both range far out to sea.

Unshaken in his belief that his bird sightings were navigational clues, Columbus on October 7 altered the course of his vessels based on his observation of large flocks of birds coming from the north to the southwest. From his prior experience with Portuguese sailors, Columbus knew that coastal birds often showed the way to land. But to his dismay, this observation also proved unreliable. Only the persistent north-east trade winds would prove reliable enough to carry him to his final destination of the island of San Salvador.

Wings Aloft and Below

While Columbus may have been disappointed that the seabirds he observed were not harbingers of terra firma, he cannot be faulted for attempting to identify and study various seabirds, all for the culmination of a successful voyage. But what exactly is a seabird? Surprisingly, there is no strict definition in the scientific literature of the term "seabird" or any defining traits or behavior patterns that clearly distinguish seabirds from other bird species. Probably the best we can say is that seabirds refer specifically to those that spend a significant part of their lives in marine environments. But even this statement falls far short, for there is marked variation among seabirds with regard to their association with or adaptation to the oceanic realm. Some birds, such as the western gull—the most common and ubiquitous gull along the California coast—are only marginally associated with the ocean, content to live and breed on the land. Away from the coast, they are relatively rare. If Columbus had observed birds falling into this category, it would have been a sure sign that land was near. In contrast, coastal seabirds, including cormorants and pelicans, never wander more than a few hours' flight from land—more or less found in shallow waters over the continental shelf. Though they may spend considerable time foraging for prey over the water, they typically return to shore daily to roost, preen, and dry their feathers. Oceanic seabirds, such as storm petrels and albatrosses, are highly pelagic in nature, fully adapted to life on the high seas, and may spend much of their lives far from the shore. They would avoid contact with land completely except during one critical period of their lives—breeding season, during which they crowd onto some of the most remote and severe pieces of real estate they can find. But as soon as their young are developed, all abandon the land to return to the open sea. Seabirds, of which there are about 250 to 300 species, comprise only 3 percent of the world's approximately 9,700 bird species that dwell on land. Ornithologists believe that great disparity in the number of seabird species compared with land birds can be attributed to the availability of niches. Although only 30 percent of our planet is comprised of land, the suitability and abundance of habitable space for birds is greater on land than in the marine environment, where niches are at a premium owing to intraspecific and interspecific competition. The storm petrel, for example, nests in crevices and burrows, sometimes shared with other seabirds or rabbits. It cannot survive on islands that have been overrun by introduced land mammals, such as rats and cats. Ornithologists recognize fifteen distinct families of marine birds, with the number of species in these families continually being updated.

The history of the modern seabird is a long one, dating back to the late Cretaceous Period (65 million years ago). One of the earliest species is *Hesperornis regalis*, a flightless loonlike bird. Outfitted with large web-like feet and a beak filled with sharp teeth, it was an efficient under-water predator, feeding mainly on small fish. While most paleontologists believe that *Hesperornis* left no descendants, Angela Milner and Stig Walsh, paleontologists at the Natural History Museum, believe that the birds of today are the direct descendants of the Cretaceous-Tertiary mass extinction event—a cataclysmic upheaval that wiped out more than three-quarters of the earth's plants and animals. But what allowed some species to survive and others, like the dinosaurs, to perish? Milner and Walsh have put forth the hypothesis that the ancestors of today's birds may have survived because of their bird brains. Analysis of computer tomography scans of two fossilized seabird brains yielded a surprising result: the brains were much more developed than previously thought, very similar to those of living birds. Even 55 million years ago, the avian brain was already modern in its complexity. Milner and Walsh came to the conclusion that birds with larger, complex brains are more flexible in their behavior, allowing them to survive when they are thrust into a rapidly changing environment.

Though modern seabirds vary greatly in lifestyle and behavior, they are examples of convergent evolution in action, having developed gradually a number of physiological and morphological adaptations that are very similar to those acquired by distant, unrelated species. In general, most seabirds are long lived—some over fifty years old—and many do not start breeding until they are five to seven years old. The latter characteristic may be an evolutionary strategy to cope with the highly variable ocean environment they often encounter in their foraging ventures. The extended time gives the birds years to learn how to find prey before breeding. And seabirds as a group have become very proficient at locating and capturing prey. Michael Brooke of Cambridge University has estimated that the annual food consumption of all the world's sea-

The Geologic Time Spiral—A Path to the Past

FIGURE 3 The Geologic Time Spiral—A Path to the Past. Source: Joseph Graham, William Newman, and John Stacy, 2008, version 1.1, U.S. Geological Survey, General Information (also available online at http://pubs.usgs.gov/gip/2008/58/)

? that's very little

birds is on the order of seventy-seven tons. To put that number in perspective, the total is similar to the global fish landings.

Unfortunately, humans often perceive seabirds differently, generally on a lower level, from other sea creatures. Megafauna, such as sharks, whales, and seals, may garner the most attention because of their size, feeding habits, and close affinity to the sea. In contrast, seabirds are viewed as aloof creatures, spending significant amounts of time flying above the ocean surface and nesting on remote islands or rocky outcrops, far from human contact. But regardless of the perception, because seabirds do derive their food from the sea, they are undeniably an important cog in the complex marine food web.

As marine predators, seabirds often occupy the upper trophic level in ocean food webs and tend to be omnivorous in their food selection, readily consuming a variety of prey items. But to a great extent, the size of the prey that can be consumed is closely related to the body size of the species. For example, storm petrels, which weigh only a few ounces, feed almost exclusively on zooplankton and small fish. At the other end of spectrum, albatrosses, weighing almost thirty pounds, feed higher on the food chain, selecting large fish and squid.

Cormorants, heavy-bodied birds, feed ravenously on fish and small eels, consuming daily about 25 percent of their body weight in food. "Cormorant" is a contraction derived from the Latin *Corvus marinus*, "sea raven." Indeed, "sea raven" was the usual term for cormorants in Germanic languages until after the Middle Ages.

The proverbial voracity of this bird came to be associated in literature with greed and malice. In Book IV of *Paradise Lost*, the seventeenth-century English poet John Milton portrays Satan as a cormorant perched in the Tree of Life in the Garden of Eden. Milton's view of the devil was an inspired transmogrification of Satan from his traditional guise, a serpent, into a black bird with a long, slender neck and an established reputation for gluttony. Shakespeare, who was an astute observer of nature and incorporated birds into many of his writings, used the word "cormorant" as synonymous with voracious, employing the word in *Coriolanus*, "the cormorant belly," and in *Love's Labour's Lost*, "cormorant devouring time."

Probably less is known about seabirds than almost any other major group of bird families, but long-term observations of their foraging habits have yielded basic feeding strategies: surface feeding, plunge diving, pilfering of food, and pursuit diving. Surface feeders tend to be

Black cormorant (sakepaint/Shutterstock.com)

the most acrobatic seabirds, hovering or gliding just above the water's surface. Storm petrels, for example, appear to lightly dance on the water surface with their long legs as they pick up small planktonic organisms, which are concentrated in converging water masses. Petrels are able to hover over the water either by rapidly beating their wings or using the wind to remain stationary. These highly maneuverable birds derive their name, "petrel," from a French version of St. Peter, who, in biblical tradition, was said to have walked on water during a storm, spurred on by the urgings of Jesus to have faith in him. These tiny birds are often at the mercy of an approaching storm, easily buffeted about by the rising winds. Early mariners viewed their frantic activity, swooping and diving just above the cresting waves, as a harbinger of stormy weather. Lore has it that sailors often called them "Mother Carey's chickens," in reference to the Virgin Mary (*Mater cara*), to whom the protection of sailors was commended in times of need.

Frigatebirds, in contrast, can glide effortlessly just above the water surface, aided by their long wing span of almost seven feet and extremely light body. (Their feathers reputedly weigh more than their skeletons.) They snatch prey, including fish, squid, and even small turtles, with their long, hooked beaks as they soar over the ocean surface. Having the largest wingspan to body weight ratio of any seabird, frigatebirds are able to stay aloft for days at a time, a significant advan-

Storm petrel (BMJ/Shutterstock.com)

tage in the search for widely dispersed prey. But to reduce wing loading, they have sacrificed leg muscles, rendering their feet useless. They cannot land on the water, cannot swim, and most assuredly cannot walk. Their sharp-clawed toes are strictly designed for latching on to a branch or any perch on which they can attain purchase. Frigatebirds are biologically designed to fly and catch food, and that is about it.

Plunge divers are able to take fast-moving prey by diving from heights into the water below. The momentum of the dive combats the natural buoyancy provided by air trapped in the bird's plumage. In general, plunge diving is probably the most specialized method of hunting employed by seabirds and in some cases, as with the brown pelican, may take several years to fully develop. The technique, once mastered, allows the birds to use less energy in pursuing prey than other feeding methods and affords them the opportunity to exploit more widely distributed food resources.

Successful foraging by plunge divers is dependent upon the detectability as well as the abundance of prey items. In the 1970s, noted ornithologist David Ainley hypothesized that water clarity was an essential factor in determining the success rate of these species: the less trans-

Great frigatebird (Elisabeth and I/Shutterstock.com)

parent the more difficult for the bird to locate food. Seawater transparency is known to affect the "encounter distance" at which visually dependent hunters, such as seabirds, fish, and some marine mammals, can detect prey. But in a detailed study of twelve species of seabirds that regularly use plunge diving to obtain food, Christopher Haney and Amy Stone concluded that water clarity was not a major influence on the foraging tactics of aerial seabirds. Interestingly, five of the twelve plunge-diving species were significantly more prevalent in turbid water. Another study of common murres, which when stationary look very much like penguins on account of their black-and-white coloration, yielded the surprising result that, though this species is considered a visual diurnal predator, murres frequently dive at night. The results from both research projects raise questions about the strategies and mechanisms birds use to find prey under very low light conditions. Hypotheses include the birds employing close-range visual or nonvisual cues to catch randomly encountered prey.

While the above studies were confined to the waters off the southeastern coast of the United States and Newfoundland, Canada, respectively, they have wide implications for the California Current Ecosystem. Analysis of a sixty-year Secchi disk (a black-and-white disk that is lowered by hand into the water to the depth at which it vanishes from sight) data set by biologists Dag Aksnes and Mark Ohman showed the transparency of the California Current has undergone a slow, but steady, decrease over this period. The researchers suggested that the decrease in water clarity was consistent with a doubling in nutrient upwelling that may have initiated light-absorbing planktonic blooms.

The brown pelican (*Pelecanus occidentalis*), entirely coastal, rarely venturing far out to sea, is the premier plunge diver, diving into the water from 50 feet or more above the waves. Upon spotting prey with its keen eyesight, the pelican thrusts its head and long bill downward, angling its wings backward to create a streamlined form. Like an Olympic diver, it corkscrews through the air, its impact cushioned by air sacs under its skin. Under the water, the pelican's bill and large pouch serve as an oversized net, holding up to three gallons of water and prey items, such as sardines and anchovies. Upon surfacing, the pelican allows the water to strain through its bill before thrusting its head back to swallow its catch.

While the water is straining, some opportunistic gulls may hover above the pelican or even sit on its bill, hoping that a food morsel slips out of the bill. The gull's audacious feeding behavior, requiring stealth and cunning, is known as kleptoparasitism, literally, parasitism by theft. As well as being the perpetrators of kleptoparasitism, gulls, generally small ones, are often victims of this act, particularly during the breeding season, by members of their own species.

Known for its brazen behavior in chasing and attacking other birds, including boobies, shearwaters, and tropicbirds, to steal their food, the frigatebird is often colloquially known as "man-o'-war bird" and "pirate of the sea." Even as far back as 1738, the English naturalist and watercolor illustrator Eleazar Albin associated the bird's name with that of frigate, a sleek ship often used for piracy. Often during seabird nesting season, frigatebirds will soar gracefully in the air currents above the colonies, patiently waiting for a parent to return with food for its young. Upon sighting the unsuspecting victim, the frigatebird, using its superior speed and agility, swoops down and gives chase to the parent bird, harassing it—frigatebirds have been known to seize tropicbirds by their long tail plumage—until the overwhelmed bird regurgitates its catch. The acrobatic frigatebird then plucks this prized bolus from the water surface or even catches it in midair. Although frigatebirds have acquired the reputation as fearless aerial pirates, kleptoparasitism is not perceived as a significant part of their diet. A study of a particular species of frigatebird, the great frigatebird (*Fregata minon*), estimated that it could obtain at most 40 percent of its food by stealing and on average only 5 percent.

While Columbus and most seamen generally welcomed the sight of birds as a good omen, the brown booby (*Sula leucogaster*), a large, tropical seabird, has often caused angst among the angling community—repeatedly attacking the trolled lures and baits of the anglers. But why? Let's take a look at the nature of this bird. The name "booby" most likely comes from the Spanish slang term *bobo*, meaning "stupid," because of the birds' fateful mistake of landing onboard sailing ships, where they were easily captured and killed for food. Though they are excellent plunge divers and are armed with straight, sharp bills that can easily dispose of unsuspecting prey, boobies are often content to sit on the water, hoping that a meal will come their way. Seemingly too lazy to secure their own food, brown boobies sometimes resort to outright stealing—a view shared by many anglers. Historically, the northern extent of the brown booby in the Pacific has been limited to the Baja California peninsula, but recent sightings of the birds along the California coast may be signaling a northward expansion of their range, possibly due to a warming climate that is more to their liking.

In contrast with frigatebirds that do not readily take to spending time in the ocean because of their inability to swim and their lack of layers of oily fat to ward off the cold water, pursuit divers—those that dive from the ocean surface to feed underwater—are at home on the rolling sea. Either wings, as used by auks or diving petrels, or webbed feet, as employed by cormorants and loons, provide the propulsion necessary to chase down fleeing prey.

One distinctive member of this diving fraternity is the tufted puffin (*Fratercula cirrhata*), a medium-sized seabird in the auk (alcid) group, which also includes the common murre, rhinoceros auklet, and black guillemot. Though puffins are about the size of pigeons, they weigh about twice as much, with an average weight of about two pounds. The adult in the breeding plumage is unmistakable, being almost entirely black, except for a white face and vibrant orange beak and feet. Its genus name *Fratercula* comes from the Latin *fratercula*, literally meaning "friar," a reference to its black-and-white coloration that resembles monastic robes. These bright breeding colors and its parrotlike beak have earned the puffin the endearing nickname "parrot of the sea."

Being plump with short wings, the puffin looks like a flying cigar in the air. The name puffin—"puffed" as in swollen or bloated—is thought

Tufted puffin (Robert L. Kothenbeutel/Shutterstock.com)

to have originally been applied to the fatty, salted meat of the Manx shearwater (*Puffinus puffinus*), a totally unrelated species, which in the middle of the seventeenth century was known as "Manks Puffin." The word "puffin" is an old Anglo-Norman word, *pophyn* or *poffin*, which was the name for cured bird carcasses. The Atlantic puffin, one of three species of puffins, ultimately acquired the name, though much later on, when it was thought that all web-footed birds were more directly related to one another.

When out at sea, where they mainly lead a solitary existence, puffins bob buoyantly among the waves, using powerful thrusts of their webbed feet to propel themselves through the water. Dipping their heads below the water surface, they scan the depths for suitable prey. Upon spotting food, they immediately dive, using their wings to stroke underwater with a flying motion. During the dive, the wings, which are partially folded to decrease drag, undergo a rapid rotary motion. The upstroke portion of the wings' movement generates a forward force, or thrust, to propel the birds rapidly downward to depths of 200 feet. As the submerged puffins "fly" through the water, they deftly use their large, webbed feet as rudders. Upon capturing their prey—generally, small fish such as herrings and sardines—puffins surface with as many as five to twenty fish held crosswise in their bills, a feat that to this day confounds biologists as to how they do so.

Not to be overshadowed by puffins as divers are the cormorants— excellent divers because their plumage is not water resistant like that of most other seabirds. They lack the oil gland used for waterproofing, thus enabling them to sink and dive deep. In addition, the cormorant's skeletal structure is dense, with bones that are solid, not hollow as typical of other species. Couple the above morphological advantages with strong leg muscles and large feet for propulsion, and cormorants can easily reach deep-dwelling prey, which they capture with eye-blinking speed. After snaring the unsuspecting fish with its hooked beak, this diving projectile quickly surfaces to consume its meal. With a dexterous flip of the fish in the air, the prey is positioned—fins back—for a final plummet down the bird's long neck.

But the cost of the meal to the cormorant is a waterlogged body riding low in the water, with only its long neck visible, like a periscope scanning the surroundings. Ultimately, the cormorant must exit the water to dry its plumage. Flapping wildly, the heavy cormorant struggles mightily to the first available perch. While graceful and sleek in the water, the cormorant moves clumsily over land, ultimately reaching a suitable site where it can spread its wings to dry—a pose that was seen as representing the cross in medieval ornamentation and heraldry, where images of cormorants were often routinely displayed. Cormorants often select a perch such that their back is toward the sun. In this position, with raised wings, they expose a black spot that absorbs the sun's heat, resulting in an increase in their core temperature.

The fishing prowess of the cormorant has long been recognized in Far East society. For hundreds of years, Japanese and Chinese fishermen have used trained cormorants to catch fish for them. To control the bird, fishermen tie a snare or leash around the base of the bird's throat. This technique prevents the bird from swallowing larger fish, which are lodged in its throat, but allows it to swallow the smaller specimens. Upon surfacing, the handler forces the cormorant to regurgitate its catch. Though cormorant fishing was once a stable commercial enterprise, it is mainly practiced today for tourists.

The adaptation to diving has left the tufted puffin, cormorant, and other efficient pursuit divers with less agility to fly. They are unable to glide and must rapidly flap their wings to stay airborne—a high energy expenditure maneuver. Since air and water have drastically different physical properties, particularly with regard to density, birds capable of moving through both media face trade-offs regarding their performance in each. Efficient flight and diving require contrasting morphological and physiological demands.

Species that exhibit great diving prowess have a relatively large body, enabling them to store large amounts of oxygen that will be metabolized during extended dives. In contrast, a large body mass is incompatible with flight because of the power-to-weight ratio. Musculature is another variable that enters into the flight-versus-diving equation. The muscles that generate flight in cormorants, for example, are small, only 17 percent of body mass, compared with other flying birds, limiting the power necessary for flight. But these foot-propelled divers have large leg muscles that can thrust them to great depths. (An extreme example of the compromise between flight and diving is found in the flightless Galapagos cormorant (*Phalacrocorax harrisi*), endowed with tiny, scruffy wings. But the loss of flight has prompted the evolution of enhanced diving capabilities in this species.) Finally, short wings are advantageous for diving because they minimize the air volume in the feathers, hence mitigating positive buoyancy, and decrease water drag because of less surface area. Wing-propelled swimmers, including auks, diving petrels, and some shearwaters, have developed into top-notch divers on account of evolutionary pressures.

Some researchers, such as Yuuki Watanabe and his colleagues, believe that the flight-versus-dive argument may have been a major factor in determining the three-dimensional foraging range of seabirds. Poor fliers that have a limited horizontal range to find food may compensate for this drawback by extending the vertical foraging component through deep dives. If prey items can be located at depth, the birds can stay in relatively close proximity to land, limiting their time away from their nests and their hungry offspring. On the flip side, aerial seabirds can cover large areas, a distinct advantage when the food supply is spatially patchy.

The acknowledged kings of flight are the albatrosses, with the wandering albatross (*Diomedea exulans*) endowed with the largest wingspan of any extant bird, reaching up to twelve feet. Since the late nineteenth century, ornithologists have puzzled over how this large bird can effortlessly soar for as long and far as it does without exerting any significant energy through flapping. Spending weeks, even months, at sea, without returning to land, the wandering albatross minimizes its energy expenditure through prolonged and sustained gliding. They even sleep while on the wing. While many birds use thermal updrafts to stay aloft, albatrosses depend upon the wind or, to be more exact, the vertical wind gradient, commonly referred to as "wind shear." It is no coincidence that albatrosses make their homes in very windy environments, for without the wind they could never get airborne and stay aloft. And the stronger the wind, the happier they are. In January 1832, Charles Darwin, while sailing aboard the *Beagle*, encountered a violent storm off Cape Horn—a region long feared by mariners for its unpredictable weather and tumultuous seas. As the captain and his crew struggled to maintain the *Beagle* on its course, fearing that the howling winds might sink the vessel, Darwin spotted an albatross above the ship's mast, appearing unperturbed by its chaotic surroundings as noted by Darwin in his journal: "Whilst we were heavily labouring, it was curious to see how the Albatross with its widely expanded wings, glided right up the wind." Even four years later, Darwin reflected that the albatross flies "as if the storm was (its) proper sphere." Blessed with a dense layer of feathers— armor against the sun, howling wind, and frigid seas—the albatross is at home in any tempest.

Watch an albatross in flight, and what strikes you is its repetitive pattern of climbs, dips, and turns, not a straight line of flight. This type of unflapping flight, which at first glance may seem bizarre, is known as dynamic soaring, whereby the albatross extracts energy from the wind field, enabling the bird to fly in any direction, even against the wind, without exerting any energy. Dynamic soaring can be broken down into four distinct components: beginning near the surface, the bird climbs into the wind; at the apex of its ascent, the albatross turns from windward to leeward; the albatross then undergoes a rapid descent to almost the sea surface; and finally, it again turns into the wind. Free as the breeze, the albatross repeats the cycle over and over again, allowing it to cover almost 600 miles in one day, without flapping its wings even once. But the underlying mechanics of dynamic soaring and the aerodynamic forces acting on the bird, which have long been recognized as being quite complex, have relatively recently come to light. Johannes Traugott and his team from the University of Munich, by combining computer modeling and GPS tracking of birds, were able to determine that wind shear coupled with the birds angling their wings into the wind generate significant lift during the ascent stage of the cycle, enabling the birds to

Laysan albatross (feathercollector/Shutterstock.com)

soar to heights of thirty to fifty feet, where the wind speed is considerably stronger than near the sea surface. The birds then reverse course, gliding with the strong wind to gain an additional speed boost as they swoop downward to the sea waves. Near the surface, in the realm of weaker winds, they turn back up again, in the process initiating a braking effect. Over the whole cycle, the birds generate enough propulsion to overcome, albeit barely, the negative effect of drag. As long as the birds religiously maintain their ritual of ascents and descents, they can fly for free. But while the albatross is at home in the wild, wide seas, terra firma presents its own challenges, as it does for other premier seabird aerialists. To become airborne, particularly without wind, an albatross needs a lot of running room—about half the length of a football field. Wildly flapping its outstretched wings as it runs, it offers up a comical picture in sharp contrast to its graceful silhouette against the sky.

Albatrosses are members of the biological order Procellariiformes, which also includes shearwaters and various types of petrels. One unifying characteristic of all Procellariiformes is their conspicuous tubular nasal passages, which are located on the upper bill. Taxonomists have used this feature to give this order its alternative name Tubinares, meaning "tube nosed." These pelagic-feeding seabirds use these passages for smelling, which aids in locating patchily distributed prey at sea. Along with being masters of soaring, albatrosses, of all the seabirds, rank the highest in the olfactory department—able to detect their dinner from as far away as twelve miles. While most seabirds use visual cues in their foraging endeavors, albatrosses depend upon their keen sense of smell in locating approximately 50 percent of their prey. Albatrosses often use the upwind component of their flight pattern to pick up the scent of prey that is being carried by the wind. Upon detecting prey, albatrosses alter their yo-yo search pattern, preferring to zigzag upwind in the direction of the faint scent signal, much like a dog following a trail.

As if out of nowhere, following its nose, an albatross may appear at the stern of a ship, hoping for a scrap of fish, bait, or some other offal that may ultimately be discarded. As long as these food offerings are relatively steady, an albatross will follow a ship for days.

Being followed by an albatross was originally viewed by early mariners as a sign of good luck, as if the bird were spreading its majestic wings over the ship in a benevolent gesture of protection. But in a bizarre contortion of logic, superstitious seamen came to see the albatross in a different light: a bad omen, which was only offering to keep the vessel from harm because the ship would soon be in dire *need* of protection from the wrath of Mother Nature. These opposing views of the albatross—a harbinger of doom or a protector of mariners—are at the heart of Samuel Taylor Coleridge's poem *The Rime of the Ancient Mariner* (1798). In the poem, a mariner shoots with a crossbow an albatross that has been following his ship. His crewmen initially fear that this wanton act of killing the bird will curse the ship and are angry with the mariner. But as the ship encounters fairer weather, the sailors have a change of heart:

> 'Twas right, said they, such birds to slay
> that bring the fog and mist.

As fate would have it, the ship becomes becalmed, with nary a breath of the south wind that the albatross had brought. And so, stranded for days on the vast sea and tormented by thirst, the sailors punish the man by forcing him to wear the dead albatross around his neck—a burden the mariner is to bear indefinitely. As Coleridge relates, all on board perish on account of dehydration, unable to take advantage of the seawater surrounding them:

> Water, water every where,
> Nor any drop to drink . . .
> And every tongue, through utter drought,
> Was withered at the root . . .
> With throats unslaked, with black lips baked,
> We could nor laugh nor wail.

Albatrosses, in contrast with humans, can readily drink ocean water. Blessed with an enlarged salt gland just behind its eye sockets, the albatross emits a saline solution that readily drips out of its nose. Other members of the Procellariiformes, such as petrels, forcibly eject the salty mixture out their nose.

Seabirds of the California Current Ecosystem

Every year, millions of seabirds utilize the highly productive but variable California Current Ecosystem. Some are year-round residents, such as gulls and pelicans, content to both breed and forage in the coastal environs. Others are transient. The short-tailed albatross (*Phoebastria albatrus*), for example, flies from its breeding grounds on the Torishima and Senkaku Islands of Japan to feed within the California Current—a journey of thousands of miles across open ocean. Not to be outdone by its albatross cousin, the sooty shearwater (*Puffinus griseus*)—a bird considerably smaller in size than an albatross—is the migratory champ, flying nearly 20,000 miles from New Zealand to the North Pacific every summer in the search for food. During its migration, the shearwater exhibits the typical "shearing" flight pattern of its genus—dipping from side to side on stiff, straight wings, the wingtips almost touching the wave tops as if to "shear" off the crest.

Extensive surveys of the seabird population of the California Current Ecosystem show that three of the four major orders of seabirds are present at least during some portion of the year: the aforementioned Procellariiformes (albatrosses, shearwaters, petrels), Pelecaniformes (pelicans, boobies, cormorants, frigate birds), and Charadriiformes (gulls, terns, auks). (Absent is the order Sphenisciformes that includes penguins.) In all, 150 species have been observed and studied.

The aggregation of seabirds within the California Current is not uniform; rather, specific locales have been shown to have large concentrations of birds. These oases of life must satisfy the birds' basic biological needs: suitable nesting and breeding sites, relative isolation from predators, including humans, and proximity to prey.

One particular "hotspot" is the Farallon Islands, small, rocky islands anchored on the relatively shallow continental shelf. Indeed, the word "farallon" is Spanish for "rock rising from the sea," and these pinnacles, which comprise more than 100 acres, were first seen by the Native Americans of the area, the Miwoks, who referred to the Farallones as the "islands of the dead," believing they visited these distant islands in spirit only. But these forbidding knots of rocks are far from being lifeless, supporting a spectacular assemblage of wildlife, including a gathering of birds not duplicated in many places of the world. The islands, located about twenty-eight miles from the coast of San Francisco, have the largest seabird breeding colony in the United States outside of Alaska and Hawaii. They are a refuge for approximately 25 percent of all of California's breeding seabirds, with more than 300,000 individuals of thirteen species. The English privateer and explorer Sir Francis Drake, after visiting the islands in 1579, was impressed by the cornucopia of life found there and wrote that they "held plentiful and great store[s] of seals and birds."

Large colonies of different seabirds can nest together on these islands because each species has evolved to employ a different type of habitat along the rocky cliffs and outcrops. Ashy storm petrels (*Oceanodroma homochroa*), of which half the world's population can be found on the southern Farallon Islands, nest and breed in small rock crevices of the islands' cliffs. These tiny birds, just over an ounce or so in weight, can wedge themselves into almost any sliver along the cliff face. In contrast, the pigeon guillemot (*Cepphus columba*), a medium-sized seabird, utilizes large rock crevices among the talus slopes and cliffs for its nesting sites. But the pigeon guillemot is not necessarily choosy with regard to its site, utilizing any suitable cavity, including sea caves and the abandoned burrows of other seabirds. The common murre (*Ursa aalge*) generally stakes out the sea cliff ledges, high above the ocean. They shun nesting material, content to lay their eggs on the bare rock. Their colonies are so densely packed that incubating adults actually are in contact with one another. While privacy is obviously at a premium, there is safety in numbers—the dense congregation helps protect the young

Common murre (Gabriela Insuratelu/Shutterstock.com)

from gulls, ravens, and other predators. Precarious as their sites are, murres are very successful nesters; they are the most abundant bird in the Farallon environs. Their high success rate is in part due to something as mundane as the shape of their eggs. Shaped like a pear, the single, large blue egg is pointed at one end and blunt at the other, allowing it to roll around in a circle on the ledge instead of tumbling off the edge of the cliff.

Many of these seabirds show remarkable site fidelity, returning to the same nest year after year, thus reducing the energy cost for prospecting for a new site. Once ownership is established, the nest will be protected from any rivals with great vigor. Young adults breeding for the first time often return to their natal colony, nesting close to where they were hatched. This tendency, known as philopatry, is strongly ingrained in some seabird species. A study of Laysan albatrosses (*Phoebastria immutabilis*) found that the average distance between their original hatching site and their subsequent nesting sites was only about sixty feet. And the champion "homer" may be a storm petrel that was caught as an adult only six feet from its natal site.

Successful breeding includes being able not only to stake out and keep prime nesting territory but also to find a suitable mate. It appears that some species resort to the sniff test. Sniffing out a genetically suitable mate is a well-known phenomenon in the mammalian world. But until recently, the prevalent thought among researchers was that birds relied primarily on vision and sound when choosing a partner. Now, new evidence shows that tube-nosed seabirds, in particular storm petrels, are able to pick out their relatives from smell alone. For years, researchers have known that captured petrels often emitted a musky scent but didn't think much about it—attributing the smell as a response to the stress of capture. The reasoning goes that if the storm petrels are able to distinguish a relative from an unrelated bird—a potential mate—then ideally this behavior prevents the birds from accidentally inbreeding. The fact that petrels use odors may also explain how these birds manage to return to the large family colony after prolonged foraging outings and are able to find their nests.

Once a bird finds a suitable site and mate, egg laying commences, and after an incubation period of approximately four to six weeks, depending upon the species, a ravenous chick emerges. Feeding styles depend upon the seabird. Puffins, common murres, and murrelets, for example, directly transfer their catch, which is held in their bills, to the waiting maws of the chicks. In contrast, regurgitation of undigested prey items is quite common in many seabird species. During their foraging outings, gulls may swallow a whole fish and upon returning to their nest regurgitate this food to their chicks. Most Procellariiformes also regurgitate stomach oil, an energy-rich food source created from digested prey. For small birds, such as the storm petrels, this is a distinct advantage since the petrels can maximize the amount of energy their chicks receive during their single visit to the nest during a twenty-four-hour period.

The nesting period of young seabirds is generally much longer than that of terrestrial species, lasting between fifty and seventy days. But with albatrosses it can take up to six months before the chicks are ready to leave the nest. While some chicks are given gentle encouragement to abandon the nest after they have fledged, the male common murre takes a different approach, swimming below the cliff and calling out to the chick. Even though it lacks the ability to fly, the chick hurls itself off the cliff edge, dropping 800 to 1,000 feet into the ocean, where it swims out to its parent.

Upon fledging it is not unusual for a seabird chick to remain with its parents for several months at sea. The brave common murre chick stays with its parent, which continues to care for and feed it until it is ready to fly. Maybe not as fortunate but ready to fend for themselves are albatross chicks, who fledge on their own and receive no further parental help. It is not unusual for adult albatrosses to return to the nest, unaware their chick has left. But once at sea, the juvenile birds rely on an innate migration behavior—a genetically coded navigation route—which aids them in their solo journey.

Visiting the Homes of Seabirds

While the Farallones may have been inaccessible to the Miwoks—their fragile, reed-built boats incapable of making it through the rough seas to the islands—a burgeoning ecotourism industry offers a variety of bird-watching expeditions during the spring and summer breeding season. (The islands themselves are strictly off limits to all except for a handful of researchers and observers from the Point Reyes Bird Observatory.) On these excursions be prepared for the unpredictable. On any given day in the Gulf of the Farallones, visibility may be severely hampered by the onset of fog, initiated by the flow of warm air over the relatively cooler coastal water. The fog quickly envelops the Farallones, cloaking them in a thick grayish blanket, making bird watching nothing more than an afterthought. But on those bluebird days, when the islands can be closely approached, your senses are overwhelmed by the sight, sounds, and, yes, smells of hundreds upon hundreds of birds. Some are frantically flapping their wings, some are effortlessly gliding, and some are diving—each driven by a sense of urgency to locate and catch food for their young.

On any single trip, even if weather and sea conditions are optimum, not every species of nesting bird will come into view. Some species, like the Leach's storm petrel (*Oceanodroma leucorhoa*) and the ashy storm petrel, are masters of stealth, only tending to their nests at night and content to the hunt the pelagic waters during the day. Even if you possess keen eyesight, it will always be easier to see an albatross, soaring with its seven-foot wingspan, than a tiny petrel, flying away from the boat. Couple the above with the dark, drab plumage of most seabirds, and seabird viewing can be quite a challenging activity.

But the high biological productivity of the surrounding waters—the result of the complex interplay of the Farallones' underwater topography and current flow—attract other species in addition to those that breed there. These migratory visitors, including several species of shear-

Habitat and Range of Selected Seabirds

Species	Habitat	Range
Ashy storm petrel	Oceanic	Southern to Central California
Black-footed albatross	Oceanic	Central California to British Columbia
Brandt's cormorant	Coastal	Southern to Northern California
Brown pelican	Coastal	Southern California
Common murre	Continental shelf	Central California to Washington
Magnificent frigatebird	Coastal	Baja California to Central California
Rhinoceros auklet	Oceanic	Southern California to British Columbia
Sooty shearwater	Oceanic	Southern California to British Columbia
Tufted puffin	Oceanic	Central California to Alaska
Western gull	Coastal	Southern California to Washington

waters, murrelets, and albatrosses, have traveled, in some cases great distances, to partake of the cornucopia of organisms that are found in the sea.

While it is truly exciting to view a flock of birds streaming past the boat (if one sees a string of black-and-white birds flying swiftly across the sea surface, chances are it is a "bazaar" or "fragrance" of common murres), the cacophony of nesting or feeding seabirds can truly be overwhelming. From small peeps and chirps to sharp, piercing calls, the cliffs are alive with the sounds of seabirds. Seabirds do not sing as do garden songbirds but vocalize through a variety of calls. Their repertoire of calls may serve several functions: impress and attract a mate, declare territorial boundaries, identify a family member, announce the presence of a predator, or convey information about food.

Among petrels, in particular, acoustic communication is an integral component of their social interactions. Their complex calls, which are used to recognize mates and other individuals, contain information about the species, sex, and identity of the caller. The harsh, wailing or squawking call of the seagull is unmistakable. The western gull (*Larus occidentalis*), whose largest single colony can be found on the southern Farallon Islands, sounds the alarm with this ear-piercing call to warn other birds of danger or even to threaten others. As the gull lowers and then raises its head, in what appears to be a triumphant gesture, this call becomes the gull's "long call," the most elaborate and individualized in its repertoire. The call of the common murre, from which it derives its name, is a distinct guttural "urr." While silent at sea, the murre is quite vocal at colony; its raucous laugh, sounding a little deranged, rings out above the incessant crashing of ocean waves against the rocky shore.

New to the world of seabirds? Have difficulty identifying a seabird by either its appearance or call? Don't fret—many tours are accompanied by naturalists trained in the science and art of observing and identifying the many species that flock to the Farallones. A trip to the Farallones, rich in life both above and below the sea, is a journey to the mysterious islands known as California's Galapagos. Though lacking in size—a small fraction of the California Current realm—the Farallones cast a big ecological and environmental shadow on these Pacific waters.

Far different from the Farallones is a rocky underwater ridge, Cordell Bank, which stands sentry to these islands. Located approximately twenty miles north of the Farallones and situated at the edge of the continental shelf, this elliptically shaped bathymetric feature plunges to a depth over 12,000 feet but at its shallowest point is only 120 feet below the surface. Ninety-three million years old, this formation was not discovered until the 1850s and remained unseen by humans until 1978. But for thousands of years, scores of species have flown and swum hundreds to thousands of miles to the twenty-six-square-mile granite mass. Over this relatively shallow feature plays out a complex interaction of currents and tides that leads to nutrient upwelling. The resultant banquet of food makes Cordell Bank a major feeding destination for thousands of migratory and resident seabirds, many of which nest on the nearby Farallon Islands.

The waters of Monterey Bay, which are rich in small fish, such as sardines and anchovies, are a haven for a large number of seabirds, ninety-four species in total, of which thirty are dominant. The vast majority of these seabirds are seasonal visitors, arriving in large numbers from temperate areas of New Zealand and Chile as well as Hawaii, Mexico, and Alaska to feed in these waters.

The distribution of seabirds appears to be correlated with the physiography of this highly productive region. In particular, water depth and distance to the continental shelf break are the most critical factors determining the habitat used by seabirds. Just offshore the water depth plunges to over 6,000 feet as the result of a deep submarine crevice, the Monterey Canyon, which bisects Monterey Bay. This submarine canyon, which is comparable in depth to the Grand Canyon, begins at

Moss Landing and extends ninety-five miles seaward. These deep waters are populated with a remarkable diversity of pelagic seabirds, such as black-footed albatrosses, ashy storm petrels and Xantus murrelets during summer and fall, and with northern fulmars and black-legged kittiwakes during winter and spring. The latter two gull-like species breed far to the north in Alaska and the Canadian Arctic but migrate south to feed on the bounty of the upwelled waters.

Along the shallow continental shelf break, a relatively narrow habitat, seabird densities are also high. During the spring and summer, sooty shearwaters are the dominant species, but by winter, fulmars and gulls make their presence known to any watchful observer.

The diversity of seabirds in Monterey Bay is, in part, linked to the presence of the Monterey Canyon. Seasonal upwelling from these great depths transports nutrients to the sunlit surface, increasing the biological richness of the area to support a robust food chain, with the seabirds as one of the top-level predators. In addition, the high nutrient content within Monterey Bay can be traced to the persistent upwelling at Point Año Nuevo to the north of the bay. An upwelling filament from this promontory travels south across the mouth of the bay and then eastward into the bay, where it is often integrated into the local circulation.

Farther to the south of the Farallones, dotting the Santa Barbara Basin, are the Channel Islands—an archipelago stretching for 160 miles between San Miguel in the north and San Clemente Island in the south. While the Farallones have never been inhabited by humans, the northern Channel Islands, which now include Anacapa, San Miguel, Santa Cruz, and Santa Rosa, were settled by and home to early Native American seafarers at least 13,000 years ago. As of the present, Santa Catalina Island, part of the southern group of islands, is the only one of the eight islands with a significant permanent population.

Though the Channel Islands are considerably south of the Farallones, their common link is the California Current and the rich broth of sea life that it supports. Bathed by these fertile waters, the Channel Islands are the southern oases for seabirds within the California Current Ecosystem—providing essential nesting and feeding grounds to 99 percent of seabirds in Southern California. Twelve species of seabirds, including half of the world's population of ashy storm petrels and western gulls as well as the only breeding population of California brown pelicans, find refuge on these isolated offshore islands.

Not to be outdone by its southern counterparts, the islands, sea stacks, and waters off Washington's Olympic Coast host nearly 100 different species of marine and coastal birds. Many of the marine birds, such as shearwaters, albatrosses, and fulmars, though having large summer populations, are not local breeders but are simply here for the food. And yet some, particularly the alcids, find suitable nesting sites. The rhinoceros auklet (*Cerohinca monocerata*), a close relative of the puffin and named for its whitish "horn" that adorns its bright orange-yellow bill, burrows into any available soil and deposits its one egg underground, as deep as fifteen feet below the surface.

Coastal seabirds, while not truly pelagic, are inextricably woven into the fabric of the ocean tapestry. Some migrating ones congregate in sheltered areas where food is abundant to fuel their journeys. Others, like loons and grebes, abandon their freshwater breeding grounds in late summer and take to the sea during the winter. The common loon (*Gavia immer*) is actually more suited to the sea than the land, coming ashore only to nest. As with other plunge divers, the loon is a powerful, agile diver that preys upon small fish in fast underwater chases. Its pointy bill, which it uses to stab and grasp its food, is no match for fleeing prey.

Researchers along this stretch of the Pacific coast are using a novel approach to monitor and study the seabird colonies along the coastline and the remote offshore islands—unmanned drones. Launched from any location on land or at sea from a boat, these thirteen-pound aircraft can fly up to two hours and cover an area of about fifty square miles, making them more efficient for seabird surveys than traditional ground-based observations. Another plus is that the drones, which fly more slowly and quietly than other aircraft, will not appreciably disturb nesting birds. Seabirds will often abandon their nests if they feel threatened or provoked. A new day may be dawning in how data are gathered about seabirds—an important step in assuring protection for these avian residents of and visitors to the California Current Ecosystem.

Seabird Threats: Past, Present, and Future

While hope about the long-term vitality of the Pacific seabird population springs from the dogged efforts of legions of dedicated conservationists, researchers, and concerned citizens, if history is any lesson, any newfound optimism may need to be tempered. When humans seek out answers for ecological degradation, they should start by looking in the mirror. Or as Pogo, the possum in the long-running American comic strip of the same name, slyly put it, "We have met the enemy and he is us."

From approximately 1807 to the 1830s, a band of Russian fur traders descended upon the Farallon Islands. Driven by a burgeoning market for marine goods, they relentlessly hunted elephant seals and sea lions for their oil, fur, and meat. Not content to limit their slaughter, they took birds and their eggs for food. Although history does not record their impact on the vast nesting colonies of seabirds, these species may well have been negatively affected.

But a sea change was on the horizon, one that would have dire consequences for the seabirds. These birds that over the centuries had adapted to incubating their eggs relatively free from molestation were soon to be invaded by hordes of men carrying clubs and sacks. In the 1850s, fueled by the California gold rush, the population of San Francisco boomed. Ostensibly due to the lack of chickens on the mainland, men stormed the islands to collect seabird eggs, especially those of the common murre. In 1855, the Farallon Egg Company was founded and remained in operation until 1881, during which time almost 14 million eggs were collected. The common murre population was decimated. This war of extermination was unsustainable. Alarmed by the wanton destruction of wildlife on the Farallones, the California Academy of Science in 1896 urged the state government to close these fragile outposts to further exploitation. Additional protection came from no less a conservation stalwart than President Theodore Roosevelt, who in 1909 declared the Farallones a national wildlife refuge.

Just as it appeared that the tide was turning with regard to human impact on the Farallones, another blow was struck. During the early part of the twentieth century, oil tankers routinely flushed their bilges near the islands. The logs of light keepers (a lost profession) are replete with entries concerning oiled birds on island shores.

The feathers of numerous seabirds are naturally waterproof, providing a barrier against the cold seas they inhabit. The quilt of feathers is a marvel in its construction: each feather is so aligned, overlapping each other like the shingles on a roof, that water cannot seep through the barbs and barbules that are part of the vane of each feather. Hooking together like Velcro, the barbs and barbules form a tight waterproof seal. Preening keeps the feathers supple and aligned. But when gooey oil sticks to a bird's feathers, it results in matting and separation of the feathers, severely impairing waterproofing and exposing the bird to the harsh elements. No amount of preening will help in realignment of the feathers, even though the bird's focus on restoring order to its plumage is so intense that it overrides all other natural behaviors: nesting, feeding, and avoiding predators. In its frantic attempt to remove the oil, the bird often finds itself ingesting oil, resulting in long-term damage to its internal organs. (Crude oil is a witch's brew of hydrocarbon-based chemicals.)

Although oil tankers are now prohibited from discharging oily residue into environmentally sensitive areas, a concern has arisen about the high volume of oil tanker traffic in the Gulf of the Farallones and Monterey Bay. On an annual basis, 800 to 1,000 tankers pass through these waters to enter San Francisco Bay. A spill from these behemoths would have dire consequences on the marine ecosystem. Some perspective is needed. When the *Exxon Valdez* ran aground in Alaskan waters in 1989, the oil from its ruptured hull spread rapidly and unchecked through Prince William Sound—a major nesting and breeding ground for northern seabirds. Though estimates vary about the resultant seabird mortality, the best guess, based on extensive surveys, puts the number at well over 200,000 birds, of which three of the nine impacted species have yet to fully recover: marbled murrelets, Kittlitz's murrelets, and pigeon guillemots.

While oil-coated birds make these animals the most visible victims of a spill, the smaller ocean creatures will bear the brunt of the damage. Exposed to the concoction of poisonous chemicals, phytoplankton and zooplankton perish. But with the demise of these minuscule organisms comes a tsunami of ecological disruption: food chains topple, and seabirds, those that can forage, find only barren waters. And timing is critical. A spill that coincides with seasonal upwelling events—periods of greatest plankton blooms—might limit the ability of plankton to rebound from the initial onslaught.

Though oil fouling, whether the result of natural causes or anthropogenic actions, creates great angst among the general public because of its obvious visible manifestations, probably more insidious to wildlife are a host of relatively invisible toxic chemicals awash in the seas. Insoluble in water and persistent in the environment, mercury and organochlorine compounds, such as PCBs and DDT, build up in the tissues of animals high on the food chain in a process known as bio-magnification—the greater number of trophic levels, the higher the contaminant concentration in apex predators.

DDT is a powerful insecticide; its potent ability to control pests was first recognized by the Swiss chemist Paul Hermann Müller in 1939. While widely adopted by the agricultural community after World War II, Rachel Carson's book *Silent Spring*, published in 1962, was a wakeup call detailing the impact on wildlife, particularly birds, of the indiscriminate spraying of DDT in the United States. Her passionate plea and the resultant public outcry eventually led in 1972 to a ban on the agricultural use of DDT in the United States.

And yet the poisons persist. A 2006 study, conducted by Myra Finkelstein and her colleagues at the University of California, Santa Cruz, has documented an increase in the levels of DDT and PCBs in both the black-footed albatross and the Laysan albatross over a ten-year period. But another finding was truly striking: the concentration of these compounds in black-footed albatrosses was 370 to 460 percent higher than in Laysan albatrosses. Biologically, both species are similar: they are long-lived, a life spanning decades; they consume a mixed diet of squid, fish, and fish eggs; and they occupy the same trophic level. The only variable that seemed to stand out was their respective foraging sites in the North Pacific. While both species breed at the same sites in the Hawaiian Islands, upon leaving their breeding colonies, the Laysan albatross migrates to Alaskan waters, and the black-footed albatross flies to the California Current. One theory that seemed plausible to the researchers is that California Current System has significantly higher concentrations of these contaminants because many of the countries that rim the Pacific Ocean do not regulate the manufacture and use of PCBs and DDT. Though the distribution and the transport of contaminants in the North Pacific involve processes that are still not fully elucidated, one cannot deny the interconnectedness of the seas. The vast swirling gyre of the North Pacific is the great equalizer, transporting water from one side of the Pacific to the other.

Mercury is a highly toxic element that occurs both naturally and as an introduced contaminant in the environment. But the toxicity of mercury depends upon on its specific chemical form and the route of exposure. Methylmercury, the most toxic form, results from the conversion of elemental mercury in waters by bacteria living in anoxic (total depletion of dissolved oxygen) environments. As with DDT and PCBs, this compound accumulates in the biota via the food chain. Fish-eating seabirds may ingest large amounts of methylmercury in their diet, which, though it is often difficult to prove cause and effect on account of other factors, has been linked to decreased reproductive success in some bird species. Studies of the effect of mercury on current seabird populations have led to other pathways of inquiry: assessing the spatial patterns and rates of increase of mercury contamination of ecosystems over time. Bird feathers from museum specimens have proved invaluable in assessing the mercury concentration in these birds when they were alive, even more than 150 years ago. The subsequent analyses have yielded some surprising results: mercury concentrations have steadily increased over time, and pelagic seabirds show higher increases than do coastal species. The latter finding possibly reflects the differences in prey consumed by the two subsets of seabirds.

Within the North Pacific gyre is a broad area, stretching thousands of miles, of light winds and weak currents but also a natural gathering point for planktonic organisms, seaweed, and other drifting objects. Matthew F. Maury offered a unique view of this oceanic feature. "Take a basin of water," he said, "and put into it some slips of wood, soapsuds and other flotsam. Then impart a circular motion to the water with a sweep of the hand and watch the result. The flotsam matter will almost directly gather into the very center of the basin, where the movement is the slightest, while the outer edge of the wheel, where the water is racing the fastest, will be left completely clear." While we can give credit to Maury for attempting to explain the concentration of material in the center of the gyre, his discussion has not held up over time.

This center of the Pacific basin is known to oceanographers as the North Pacific Subtropical Convergence Zone; it has recently come to be associated with the "Great Pacific Garbage Patch"—a collection of marine debris that has been drawn in by wind-driven Ekman transport, not the circular motion of the gyre. Though some reports about this "garbage patch" have hyped its size and nature—a floating landfill the size of Texas, with miles upon miles of bobbing plastic bottles and containers—in reality, much of the debris found in this area are small bits of plastics (microplastics) suspended throughout the water column.

Many seabirds accidentally ingest microplastics, which are often difficult to see, like shards of glass in a carpet, as they forage. While plunge divers manage to avoid the surface realms of floating plastic, some birds are not as fortunate. Surface feeders, such as albatrosses, petrels, and fulmars, which skim the water surface with their bills, can accumulate plastic in their gut. Adults, in most cases, can regurgitate the plastic particles when they feed their chicks, minimizing the impact on them, but the chicks are unable to. At present, there is no general consensus within the scientific community on the short- and long-term effects of plastic ingestion by seabird chicks, but the impacts might include laceration of the stomach lining or accumulation in the gut, leading to a false sense of fullness that can result in starvation.

Homo sapiens as a group are fallible, prone to missteps, and nowhere is this more evident than in the impact of fisheries on seabird populations. Most seabird mortalities, as with sea turtles, are the result of commercial fishing operations.

Surface-foraging birds are comfortable scavenging for dead or moribund prey, making them "conditioned" to seek out supplements to their natural diet, thus "stealing" bait from longline hooks. Many seabirds become hooked in this manner and subsequently drown. Longline fleets—which in total set more than 1 billion hooks each year—account every year for the deaths of up to 300,000 seabirds, including 100,000 albatrosses.

In the North Pacific, pelagic drift gill nets were estimated to have killed 500,000 seabirds annually, mostly migrating sooty and short-tailed shearwaters, prior to a United Nations moratorium in 1992 on the use of these nets on the high seas. Traditionally, gill nets were constructed from hemp, cotton, or multifilament nylon, which by its very nature was highly visible to seabirds. Over the last few decades, a shift occurred to the use of monofilament nylon—cheap, long lasting, and easy to use—which is essentially invisible to seabirds. These nets are particularly lethal to pursuit divers that forage below the surface. In California during the 1970s and 1980s, as fishing pressure increased and monofilament nets became more prevalent, colonies of the common guillemot—capable of reaching depths approaching 600 feet—declined rapidly. By the mid-1990s, estimates placed the death toll between 6,000 and 13,000 birds annually. With gillnetting highly restricted today in California, the common guillemot populations have rebounded. Farther north, in Washington and British Columbia, the common guillemot has not fared as well, remaining the species most frequently caught in coastal gill nets that have also indiscriminately entangled rhinoceros auklets, marbled murrelets, sooty shearwaters, and Brandt's cormorants.

To stem the decline in seabird populations, fundamental changes in fishing practices need to occur. But fishery management strategies have historically been aimed primarily at optimizing commercial fish yield to the exclusion of affording protection to other aspects of the marine ecosystem. Usually, protective measures are not addressed until after a problem has been identified, and by then, the damage may be irreversible. If mitigation measures are developed in a timely manner, they must be followed up with implementation and vigorous enforcement and, possibly most important, have the support of the fishing community to be effective.

Many fishermen find it troublesome, never mind economically disadvantageous, to deal with avian bycatch in their gear. If they can be convinced that the proposed conservation measures will not reduce their catch of target fish species, they are more likely to accept changes to their fishing practices and/or gear. Independent observers on commercial fishing vessels, for example, have documented that most seabirds—many being visually dependent hunters—are hooked when longlines are set during the day. A changeover to setting at night might be an effective mitigation approach.

The dramatic rise of industrialized fishing—large factory ships capable of catching, processing, and storing huge quantities of fish—in the 1960s, and its continuance today, has led to overfishing of many species, mostly higher-trophic-level fish, throughout the oceans. But there is growing evidence that the impacts of overharvesting are not limited to the target species but may result in structural changes to other marine communities, including those of seabirds. A novel approach to assess the nature of these changes has been developed by a research team from Michigan State University and the Smithsonian Institution: an analysis of the bone chemistry of both ancient and modern seabirds. To the scientists, the bones were a storehouse of diet and foraging information. Using the remains of Hawaiian petrels—seabirds that spend the majority of their lives searching for prey in the open Pacific—they were astonished to see a dramatic shift in their forage items. Between 4,000 and 100 years ago, petrels consumed large prey items, but after the onset of industrialized fishing, which targeted top-level organisms, their diet consisted mainly of smaller fish and other organisms. The bone record was a telling story of the human impact on open-ocean food webs.

But there may be other chapters to be added to this tale. The lack of apex fish available to fishers has led to a gradual, but steady, shift to targeting smaller species—a phenomenon known as "fishing down the food web." The industry's emphasis on harvesting lower-trophic organisms, primarily for fish meal, appears to be on a collision course with the seabird community.

Questions arise: How much food is needed by seabirds? Is there a critical level below which seabirds suffer? Answers are in short supply due in part to the dynamic nature of the birds' environment, but an extensive study of fourteen seabird species that feed mainly on small fish, such as sand lance, herring, and sardine, found that when food levels dropped below one-third of the maximum amount of food available historically, a tipping point was reached: breeding success declined. Regardless of where the seabirds occurred in the world, the effect of low fish abundance was the same. If food, however, was relatively plentiful, above the one-third critical level, the number of chicks produced was unaffected by short-term fluctuations in food availability. This massive data-mining undertaking—covering seabirds from the Arctic to the Antarctic and comprising a total of 438 years of painstaking observations—established an important benchmark, a guide for resource managers and policy makers when proposing fishing limits in order to maintain robust seabird populations. The international team of scientists, though assuming a cautious posture with regard to their findings, has adopted the mantra "one-third for the birds" as a starting point when considering broader approaches and options for ensuring marine ecosystem sustainability.

The problem of food availability for seabirds can be more acute in upwelling areas, such as the California Current System, where fishing industries capture the same fish species, sardines and anchovies, upon which seabirds feed and where El Niño events can lead to wide fluctuations in the quantity of these small pelagic fish. The 1982–83 El Niño and associated anomalous warming negatively impacted Pacific seabird populations. Effects ranged from abandonment of traditional hunting sites to nest desertion to mortality. In Oregon, the warming was pervasive, extending at least 120 miles from the coast and down to depths of more than 300 feet, and coincided with the seabird breeding season. Those seabirds that attempted to complete the nesting cycle during this period of food shortage become weak, unable to reproduce, and also possibly sacrificed future reproductive attempts.

Three species, Brandt's cormorant, the pelagic cormorant, and the common murre, had markedly reduced breeding success compared with previous years. Nest abandonment by cormorants was widespread in 1983, as these birds most likely spent more time at sea searching for new feeding opportunities. Mortalities were common, particularly for pigeon guillemots and the adult murres, which were found dead in significant numbers on the coastal beaches.

Seabirds are the "miner's canary" in the ocean, often the first to signal that there is an imbalance in the food web or a general deterioration in the health of an ecosystem. A recent study of Baja California pelicans, gulls, and terns found that the proportion of sardines and anchovies in their diet should be a red flag to the fishing industry to reduce catches in the future to prevent a collapse of the fishery. In upwelling regions, commercial interests and seabirds often compete for the same resource: small forage fish that are vulnerable to oceanographic anomalies. The relationship between the proportion of sardines in the diet of seabirds and sardine landings showed that the amount of sardines caught by seabirds decreased faster than it did for the fleet, which can maintain relatively high catches, even when seabirds have shifted to new prey items, mostly anchovies. The lesson to be learned was that even when present catches were high and showed no sign of decline, seabirds were recognizing and confronting a change in their regime, signaling that fishing pressure should be reduced. This alone should warrant our respect for the myriad seabird species with which we share this planet.

Chapter Four

ANGLERS REAP THE BOUNTY

The year was 1919. And although World War I had been over for a year, a U.S. Navy plane slowly searched the waters off San Diego for its target. On a dark, moonless night, crew members were scanning the waters for bright shimmers and crescents of light. Once sighted, the crew would transmit the coordinates to a naval base ashore, starting a chain reaction ultimately culminating in ten to fifteen pursuit vessels beginning the hunt. Once one of these vessels came upon the target, the entire fleet would enter into an attack mode, employing a number of capture techniques.

This was no search-and-destroy mission of an enemy submarine but the use of the latest twentieth-century technology to assist Southern Californian fishermen in locating schools of Pacific sardine. This silvery, tiny fish, a mere six inches in length, would be at the center of a commercial fishing enterprise that at its peak would provide jobs to thousands, even during the years of the Great Depression. But a fishery that once seemed boundless would ultimately crash, forming the backdrop of John Steinbeck's novel *Cannery Row*.

If America is the world's melting pot of many ethnicities, then probably nowhere is this truer than in the commercial fishing communities that dotted and prospered along the Pacific coast. California's commercial fishing industry, in particular, has a long, colorful history, enriched by emigrants from Europe and Asia who put down roots up and down the long coast. The ingenuity, work ethic, and skill of these fishermen and the commercial fishing industry they developed were instrumental in the growth of local economies from Crescent City in northern California to San Diego in the south.

Beginning at the turn of the twentieth century, Europeans were drawn to the fertile waters off the California coast; a new "gold rush" was starting, but one centered on fish. These hardy fishermen brought their time-tested fishing methods from their native countries and adapted them to California waters. The various nationalities settled along different parts of the coast, pulled by heritage, fishing specialty, and fish. In San Diego, Portuguese communities sprouted up to fish for tuna with line and hook. Fishermen from Italy and Yugoslavia settled the San Pedro area to net sardines and squid. Santa Barbara attracted Italians, many from Genoa, who trapped lobsters and tried their luck fishing. Sicilians, who settled in Monterey, introduced the lampara net as a means of catching sardines. The San Francisco area fleet also hailed from Sicily—crabbers and trollers who plied the waters north to Bodega Bay and south to Half Moon Bay. As early as 1870, these fishermen were providing 90 percent of all seafood consumed in San Francisco. Scandinavians were also represented, Finns and Norwegians settling in Fort Bragg, Eureka, and Crescent City.

Preceding the influx of Europeans, Asians had fished California waters as early at the mid-1800s. Chinese fishermen initiated the California squid fishery in 1863 off Monterey Bay. Later on, they were joined by Japanese, forming a successful coalition that caught, dried, and exported squid to U.S. markets until the mid-1930s.

The diversity of fishermen and their proven fishing methods, as well as the variety and abundance of fish, would propel California to the top of the U.S. commercial fishing industry during the twentieth century. New fisheries were, it appeared, springing up almost overnight.

Tuna Fishery

Probably no other single factor contributed to the explosive growth of the Southern California tuna fishery than the ability to process and can one's catch. The canning of fish, according to most sources, probably originated in France during the early part of the nineteenth century. French biochemist Nicholas Appert provided food for Napoleon Bonaparte's army, employing a relatively simple but effective technique: fresh or cooked fish that had been placed in a sealed bottle was heated in boiling water. Food treated in this manner retained its flavor and could still be consumed weeks later.

In these early years, bluefin and albacore tuna were the main species caught and canned owing to their availability and abundance in the temperate waters of the North Atlantic and the Mediterranean Sea. The canning of tuna did not become an industry in the United States until 1903, spurred on by the lack of sardines. Because of the uncertain sardine supply, Southern California fishermen turned their efforts to seeking out other fish to catch and can. Albert Halfhill was one of the first industrialists to develop a tuna canning operation, centered on albacore tuna (*Thunnus alalunga*): the California Fish Company, based in San Pedro. While trying out several ideas he had for cooking fish, Halfhill discovered the flesh of albacore turned an appealing white shade when cooked under pressure by steam. (Ultimately, albacore would be marketed and labeled as the "white-meat" tuna, a term still in vogue today.) To Halfhill's dismay, canned albacore was not, at first, popular in California, but undeterred by this setback, Halfhill shipped his product eastward, primarily to markets on the Eastern Seaboard, where it was an "instant" success primarily owing to Halfhill's tireless salesmanship. Following in the footsteps of Halfhill and buoyed by demand for his product (consumers came to appreciate the long shelf life of canned tuna and its tasty flesh), other entrepreneurs opened five new canneries between 1911 and 1912. In 1914, Frank Van Camp and his son bought the California Tuna Canning Company and changed the name to the Van Camp Seafood Company. As a marketing tactic, Van Camp employed the term "chicken of the sea" as a way to describe the tuna's taste. This strategy was so successful that it soon became the company's name. The demand for tuna skyrocketed, and Southern California ports, including San Pedro, Long Beach, and San Diego, became the epicenter of a robust tuna fishery. The albacore's relatively small size (twenty to thirty pounds) and its tendency to form large schools made it ideal for fishermen to catch—a plus in attempting to supply enough tuna to the burgeoning cannery industry.

As the albacore tuna fishery developed off Southern California, it became increasingly dominated by Japanese immigrants, many of whom settled in San Pedro. By 1914, over 150 Japanese fishermen were operating 50 of the 130 vessels that were based in San Pedro. Canners estimated that 80 to 90 percent of the albacore catch was landed by what they derisively labeled the "Japanese monopoly."

One particularly successful Japanese immigrant was Masaharu Kondo, who in 1912 founded the M. K. Fisheries Company, which operated out of San Diego. Kondo recruited experienced Japanese lobster and abalone fishermen to his growing tuna enterprise and, along with other Japanese fishermen, introduced the use of long, slender bamboo poles to catch tuna.

Each bamboo pole was equipped with a short line, wire leader, and

Albacore tuna (holbox/Shutterstock.com)

barbless hook. Bamboo-pole fishermen did not cast the line in the conventional sense but slapped the line on the water, creating a noisy popping sound that, hopefully, would catch the attention of tuna. When the tuna were biting, a routine set in: slap, pull, and haul out a fish. As Motosuke Tsuida, captain of a tuna boat, wrote, "The Japanese fishermen could catch the fish fast. Our men fished with a certain rhythm, and the fish would come off the hook in mid-air. From a distance, the tuna looked like silver petals falling from a tree."

For larger tuna, two men would work in tandem. The heavy cotton lines from their two poles would pass through a swivel, then join to a single leader and hook. For even bigger fish, three poles, sometimes four, might be needed. When a tuna struck, all the men lunged backward in unison, catapulting the great fish out of the sea and into the boat.

The tuna, which could often be spread out over a wide area, were attracted to the boat, generally, by means of chumming. Until the 1920s, tuna boats used chopped sardines for chum, which while relatively effective had a downside: its monstrous stench, particularly after baking in the sun, left many a hardened fisherman gasping for air. Upon arriving

Pole-and-line fishing (courtesy of NOAA)

on the Pacific coast, the Japanese employed live bait. The frantic splashing of the panicked bait created a more realistic setting, inciting the tuna into a feeding frenzy where they would strike at even bare hooks.

Despite the documented success of pole-and-line fishing, one canner, Frank Van Camp, was unconvinced about the efficacy of bamboo poles, arguing that the method was not efficient enough to catch all the albacore available during the fishing season. Consequently, Van Camp financed the use of an experimental net, firmly believing that nets would outperform bamboo poles in catching albacore. To Van Camp's dismay, the nets, which were extremely heavy when saturated, awkward to handle, and dangerous in rough weather, were unsuccessful in catching tuna. The editor of the *Pacific Fisherman*, writing in the 1915 issue, grudgingly acknowledged that "what the Japanese lack in strength and ability seems to be more than made up in patience and unique methods of luring the fish to take the hook." By the 1920s, the rest of the tuna fishery came to realize that the most productive way to fish for tuna was with a pole and line, which sometimes accounted for up to 60 percent of the catch. (The method remained popular for decades, but by the early 1980s, its use in the albacore fishery had declined, accounting for only 10 percent of the catch as compared with 90 percent by trolling jigs.)

The albacore tuna is a highly migratory, pelagic species. These nomadic wanderers were probably first recognized by Fray Martin Sarmiento in 1757 when he wrote, "Tuna have no native country, nor lasting home. All the sea is their native country."

Albacore tuna generally arrived in Southern California waters in May, but at times their presence was not noted until July. They seemed to travel some well-defined sea highway, possibly the California Current, to reach the coast near San Diego. By September, the albacore had worked their way back south. When the fish left, the canneries had no choice but to shut down their operations.

But even with this downtime, the growth of the tuna industry would remain on an upward trajectory. By July 1914, eleven canneries were operating in Southern California, only to find the bubble burst. In October 1914, the tuna industry was set back by the unexpected "disappearance" of albacore. The canneries struggled to stay afloat. The albacore tuna shortage strained inventories and caused the young industry's woes to be the center of increased media attention because the demand for canned albacore could not be met. It should be pointed out that bluefin tuna (*Thunnus orientalis*) were also available in the Southern California waters. But three obstacles stood in the way of harvesting this fish: fishermen had not yet developed the gear to target bluefin efficiently; cooked bluefin did not have the appealing "chicken white" appearance as did albacore; and probably most important, bluefin were a prized recreational target for the powerful and influential sport-fishing community in California. Challenging this group would be a grave and costly political mistake.

While tensions rose among the canneries as the older, more established companies accused the newcomers of creating a massive advertising campaign based on an unreliable supply of albacore, some cannery owners took a more pragmatic view of the problem, recognizing the need for reliable scientific studies of the spawning habits and migration of albacore tuna. These sentiments were echoed by Russell Palmer of the *Pacific Fisherman*, who in January of 1915 warned that without adequate scientific data regarding "the source of raw material, the financial confidence" in the canned tuna industry rested on shaky ground.

By the end of 1915, tuna canners had petitioned the federal government to ramp up scientific research on albacore tuna. Even the results of preliminary at-sea investigations were ambiguous; solid answers about tuna availability and sustainability could not be provided to the canners. With this great uncertainty surrounding the albacore supply, the very existence of the Southern California canned tuna industry was being threatened. Not until 1930 would the United States government make some genuine strides toward addressing the issue of sustainability of the tuna industry. Geraldine Connor of the California Division of Fisheries preached the need for an international stewardship of tuna in the eastern Pacific. Connor justified her plea by stating that it "will naturally work to the good of the industry as well as to the preservation of the fish." After years of negotiation and wrangling, the Inter-American Tropical Tuna Commission (IATTC) was founded in 1949, with the explicit mandate to conserve and manage the tuna resources of the eastern Pacific. While the original members included only Costa Rico and the United States, presently sixteen nations are part of IATTC.

The outbreak of World War I in Europe provided a unique opportunity for the fishing industry to look for new seafood options as alternatives to canned albacore. Canned sardines, in particular, became an important commodity, with new markets springing up both domestically and internationally. But around the same time, bluefin tuna came back into the picture as the result of successful experiments conducted by fishermen employing purse seine nets (a net that is drawn around a school of fish and then closed, "pursed," at the bottom by means of a line passing through rings attached along the lower edge of the net). On August 19, 1917, four purse seine boats recorded the largest catch of bluefin tuna known to the industry. Most of the seventy-five tons of tuna caught were canned by the Van Camp Seafood Company for the domestic Italian market. The success of these purse seiners was an eye opener to many in light of the failure of the same method to catch albacore. About a year later, almost 1,000 tons of bluefin tuna were landed—a record catch of fish at that time for Southern California. While peace was celebrated throughout the United States with the cessation of hostilities in Europe by the fall of 1918, the good fortunes experienced by the tuna industry would come to an abrupt end. For the next three years, the industry was beset by economic troubles, causing many tuna canneries to permanently close their doors, merge, or be acquired by other companies. To add to the misery, the albacore still had not returned; the 1920 season was a disaster, causing eight canneries to close in San Pedro and another two in San Diego. Even two years later, the tide had not turned. As one tuna cannery manager bemoaned, it was the worst season he had ever experienced in his twelve-year career. And the fishermen would soon absorb another blow. The bluefin supply, which had previously flourished during the war years, was a complete failure. Unbeknownst to these hardy fishermen, the cause for the collapse of the tuna fishery was most likely due to the fickle the nature of the California Current Ecosystem. The systematic study of the oceans and their impact on marine life was still in its infancy. Oceanographic-climatic events, such as El Niño–Southern Oscillation, were not in the common vernacular. The cannery owners, fishermen, and consumers were simply left scratching their heads as to the demise of a once-profitable enterprise.

The albacore catch continued to fluctuate during the 1920s, much to the frustration of both fishermen and canners, who depended on a steady supply of fish to meet the consumer demand. In 1925, 9,900 tons of fish were boated—an all-time high-water mark for the industry—but within a year, the catch in coastal waters had dwindled to less than 1,200 tons. The cause for the decline appears to have been a shift in albacore to offshore waters, striking fear in the fishermen that the albacore might disappear from the coast altogether. The total catch rose a bit in 1927, but the following year the fishery reached rock bottom, with less than 300 tons taken in coastal waters. During the latter part of the 1920s, it appears that the albacore fishery was undergoing a classic "boom and bust" cycle: as a fish stock increases, it results in increased landings, only to be followed by a collapse in the fishery due to overfishing.

The summer of 1929 was marked by guarded optimism by the fishermen; rumors were circulating of albacore catches—big catches—at some unspecified distant port. Fishermen were confident that the "long fins," as they were often referred to because of their oversized pectoral fins, were returning in large numbers. The sighting of huge schools of albacore fueled the fishermen's optimism. But by late August, only deep disappointment was to be found within the fishing community; the catch had not exceeded 130 tons. In the September 1929 issue of *West Coast Fisheries* magazine, George Chute wrote: "Therefore we know the end is here. And a queer sort of end it is, for it seems almost like the conclusion of a thing that never had a beginning. Rather than the end of actual fishing, it marks the end of our hopes, for all of us—as always—have hoped against hope that this year, this year, the fish would return at last."

With the lack of abundant albacore tuna supplies in Southern California and to combat the economic downturn, canners, at least those still in business, and Japanese investors focused their attention on Mexican waters, particularly those off Cabo San Lucas in southern Baja California. The Van Camp Seafood Company was one of the first canneries to realize that Cabo San Lucas had an abundant supply of yellowfin (*Thunnus albacares*) and skipjack (*Katsuwonus pelamis*) tuna and that the fishermen did not have to be limited to seasonal fishing. Advertising campaigns sprang up extolling the virtues of "light-meat" instead of "white-meat" tuna. Investments in this new fishery poured in as confidence soared in this relatively new untapped resource. These investments manifested themselves in the form of refrigerated schooners—a boon to the Southern California–based fleet that could preserve its catch on their long journey southward.

Stories circulated about the abundance and size of these warm-water tuna. While skipjack tuna are the smallest of the tuna species (18 to 22 pounds), Pacific yellowfin top out at over 300 pounds, but the same species that swims the Atlantic rarely reaches a weight of over 100 pounds.

Yellowfin tuna are one of fastest-growing tuna species, packing on weight even more quickly than the bluefin, which can reach weights over 1,000 pounds. With abundant food resources in the southern portion of the California Current, during its average eight-year life span the yellowfin tuna can efficiently convert these resources into weight gain. The sticking point appears to be what is meant by efficiency and what are its determining factors. Efficiency from a biological perspective can be viewed as the conversion of an amount of food consumed into weight gain. For most organisms, the efficiency is only a few percentage points. For argument's sake, a fish that consumes ten pounds of prey may gain only one pound—an efficiency of 10 percent. Why such a low number? There are loses, such as metabolism and waste. But energy expenditures depend upon activity and mobility. In the Atlantic, yellowfin tuna migrate from their tropical haunts to the New England area during the spring and summer, riding the warm Gulf Stream northward, which allows them to remain in their thermal comfort zone. But during their journeys, food is scarce and energy expenditures are high, thus negating any significant weight gain.

In the Pacific, underwater hills and seamounts were found to be magnets for these fish. Localized upwelling induced by these bathymetric features initiated a robust food web, which could support large numbers of prey and the predators that fed on them. Manuel Rosa, a Portuguese skipper, discovered a fishing bank southwest of Sebastian Vizcaino Bay in Baja California, which became known as "Rosa Bank." He later recalled that "the tuna were so thick that it looked as if one could have walked across the ocean on their backs. They were hungry, too, and it required very little time to fill our boat to capacity." (Even today, Rosa Bank is a prime destination for recreational anglers aboard the long-distance boats that depart from San Diego.)

With the opportunity to fish for tuna off Baja California, California fishermen gambled on building even larger, faster, and more efficient vessels as a way of increasing their profits. The mid-1920s marked the beginning of the era of the "tuna clipper." In a short time, a wide-ranging, live-bait fleet had been created, capable of traveling thousands of miles and designed specifically for fishing off the Mexican coast.

At the forefront of this surge in boat building was Manuel Medina of San Diego, who designed and built one of the first great clippers. The *Atlantic* was 112 feet long, the first tuna boat to break the 100-foot mark, and was equipped with state-of-the art refrigeration and a powerful diesel engine.

A revolution in boat building was under way, one that not only impacted American interests but would also play a central role in Japanese expansion into the Mexican tuna market. Japanese tuna fishermen designed a "new style" tuna boat, one capable of carrying enough bait, fuel, and provisions for lengthy fishing trips on the high seas, otherwise known as international waters.

These vessels were hugely successful, responsible for large yellowfin and also skipjack tuna landings in 1927, making it the first time that over 50 percent of tuna processed in California ports had been caught south of the border. The March 1928 issue of the *Pacific Fisherman* was effusive in its praise of these new "vessels that are capable of bringing in their own fish, making them independent of tenders, and vastly increasing the radius of operation that has been found to be of great advantage in developing new and productive grounds." The tidal wave of shipbuilding continued to grow into the late 1920s, with thirty-five new tuna clippers launched, seven vessels converted into seaworthy tuna boats, and ten clippers built by the Japanese to bolster their California-based tuna fleet. These new clippers expanded the boundaries of tuna fishing, reaching as far south as Ecuador—an area of operation greater than that of the continental United States. As the fleet grew, it ultimately settled into a pattern of operation based on the poorly understood movements and habits of tuna. A study by the California Bureau of Marine Fisheries reported that from November to February the fleet exploited the Galapagos Islands. (In 1929, the *Atlantic*, with Medina at the helm, sailed past Cabo San Lucas, past Cocos Island off the coast of Central America, and was the first to drop anchor in the Galapagos group.) In the spring, the boats fished the waters off Central America. When June and July rolled around, the fleet moved northward to the Gulf of California and Cabo San Lucas and by August and September had scattered along the Baja California coast, fishing the neighboring banks and islands. Through the autumn months, the fleet mostly reverted to Central America. The San Diego–based tuna fleet was now fishing almost year-round and in the process setting record catches. The canneries in San Diego had their busiest year in 1935.

As more tuna clippers sailed southward to tap into the unlimited fishing potential off Mexican shores, political issues surrounding this influx of vessels into Mexican waters arose. Mexican officials, looking to capitalize on these fishing ventures, imposed a hefty export tax on tuna. Canners and fishermen, distraught that this tax would severely cut into their profits, petitioned the U.S. government to resolve this thorny issue. On December 23, 1925, a treaty was signed to define international boundaries in relation to the enforcement of duties as well as creating an International Fisheries Commission to monitor relations between the United States and Mexico. The hope for a long-lasting agreement was abruptly shattered when several U.S. tuna vessels were seized, their crews arrested, and the owners fined. By 1927, as tensions mounted, the treaty was terminated.

Despite this political setback, the tuna industry exhibited great resiliency, showing no signs of slowing down. In fact, the industry continued to adapt and evolve, particularly with regard to the use of the gear employed on tuna vessels. In the early 1930s, the hand pulling of a seine net over a turning roller was replaced by means of a boom and a winch, dramatically reducing the amount of time and physical labor required to pull in the net.

As a result of this innovation, the years to follow would show a growth in the number of purse seiners built. Unfortunately, these vessels were not consistently successful, and many were converted to bait boats. Even a few years after World War II, when new steel-hulled vessels entered the tuna fleet as seiners, there was no guarantee that these vessels were there to stay. To make matters worse, fishery scientists concluded that there was no hard evidence that purse seining increased fish production compared with more traditional fishing methods. Interest in purse seining was ebbing quickly.

The tide turned as a result of further advancements in the technology of purse seining: the use of nylon nets and the Puretic power block. Nylon nets are vastly superior to cotton nets because they are lighter and stronger and have an additional advantage in that they do not rot—a plus in tropical conditions. The power block, invented by San Pedro fisherman Mario Puratić, is a hydraulically powered pulley attached to the end of a boom. The net is pulled upward through the block, where it then moves downward to the stern of the boat to be tended to by the fishermen. By virtually eliminating arduous manual labor, the power block made the retrieval of the net easier, faster, and safer.

But would these two innovations be the salvation of the tuna purse seine industry? The following years for the industry were rocky, marked by a number of missteps when bait boats were converted into financially expensive seiners that had only marginal success. Even big industry giants, such as Star-Kist, were not convinced that seiners were worth the risk and the expense, telling one captain of a vessel it owned that he would "starve to death" if the boat was converted to a tuna seiner.

But Captain Lou Brito, through his analysis of these past ventures into seining, viewed purse seining as the future of tuna fishing. With the support of Joseph Martinac Jr. and Nick Bogdanovich, owners of Brito's vessel the *Southern Pacific*, Brito converted the vessel into a seiner during 1958. The maiden voyage of the newly outfitted ship was wildly successful, yielding 231 tons of yellowfin tuna on a thirty-day trip. In 1959, Brito topped that amount, returning triumphantly to San Diego with 235 tons of yellowfin. Brito's huge success, achieved in a relatively short time, made him, in the minds of many, the savior of the San Diego tuna fleet. Soon to follow, the boat building floodgates would open, with over 160 new tuna seiners built in the United States between 1961 and 2000.

During the two decades after Brito's fishing ventures, other countries, including Japan, Spain, France, Taiwan, South Korea, Mexico, and Venezuela, adopted the vessel design and gear technology that had given U.S. fleets a competitive edge in the tuna fishery. From 1990 to 2000, the average annual tuna catch was about 3.4 million metric tons, an amount nowhere near approached in previous decades. And by 2003, over 500 "super" seiners—a new breed of even more technologically advanced ships—combed the tropical seas for tuna, often in complete disregard for long-term resource sustainability.

This unprecedented growth in the purse seine industry was, interestingly, predicted by Dr. Tage Skogsberg of the California State Fisheries Laboratory as far back as 1925, when he noted that the development of the California ocean fisheries would follow the European model, that is, a future predicated on developing distant fishing grounds. In his report to the California Fish and Game Commission, he stressed the need for "vessels large, speedy, and economical enough to operate with profit as far away as the Gulf of California and along the west coast of South America."

In 1960s, the Pacific purse seine fishery entered into a controversy that would continue to plague it for decades. What happened? The American tuna industry developed a new way of fishing, called "fishing on dolphin."

In the eastern tropical Pacific (ETP), large yellowfin tuna often swim together with several species of dolphins, including spinner, spotted, and common dolphins. This ecological association of tuna and dolphins is not fully understood, but clearly it had two important consequences: it was the foundation of a successful tuna fishery and resulted in the deaths of a large number of dolphins. The latter is at the heart of the controversy.

In the ETP purse seine tuna fishery, the nets may be set around schools of tuna associated with dolphins, a method commonly referred to as a "dolphin set." The dolphins are an integral part of the fishing operation. As marine mammals, the dolphins drown as the net is retrieved back to the mother ship. While the number of dolphins killed since the fishery began has been hard to determine because purse seiners often operate on the largely unregulated high seas, estimates place the number in the millions, the highest known for any fishery. When the U.S. Marine Mammal Protection Act was passed in 1972, it included provisions to reduce bycatch to insignificant levels. With the vigorous enforcement of the law and the advent of the popular "dolphin-safe tuna" campaign—tuna caught without setting on dolphins—in the 1990s, the tuna industry was pressured into making significant changes to reduce dolphin mortality. Over time, the bycatch of dolphins has plummeted significantly.

To the average consumer, the label "dolphin-safe tuna" means that no dolphins were harmed in the process of catching tuna. But in reality what it means is that one particular method cannot be employed to catch tuna in one specific place. Tuna companies have knowingly mis-led the consumer and obscured the fact that other methods of catching tuna may be more harmful than setting a net around a pod of dolphins. Environmental activists arguing for transparency in the tuna fishery have trumpeted that dolphin safe is not ocean safe. In particular, when nets are set around fish-aggregating devices, tethered buoys or floats that congregate marine organisms around them, the result is massive bycatch, including turtles, sharks, and billfish.

But what about the tuna—the albacore—that started the mad rush almost a century ago to bring tuna to worldwide markets? What is the current health of the once prosperous Southern California fishery? The most recent albacore population survey (2013), as documented by Michael Moore in the *Fishermen's News*, yielded a depressing result: an absence of albacore in Southern California waters. Most of the alba-core were found farther north, roaming the offshore waters of Oregon and Washington. In fact, it has been at least ten years since there has been enough fish to supply the Southern California fleet. The case of the "missing" tuna has left scientists searching for answers, with specula-tions swirling about, including changes in sea surface temperature and food availability. Or as Moore writes, "If an albacore swims more than 60 miles off the southern California coast and no one is looking for it, does that mean it's not there?" Good question. Over the last decade, there have been very few fishermen working Southern California waters for albacore (concentrating instead on the more profitable salmon found in northern waters). In addition, it appears that commercial fishing is losing its appeal. Back in 1968, about 2,300 boats made up the South-ern California fleet; today, that number has dwindled to between 600 and 700. While fishing was in the blood of the early immigrants—a way of life—fewer and fewer want to follow in their footsteps. A fishery that has been around for over a century will probably survive—most likely in a different form and on a smaller scale—as scientists, managers, and, yes, fishermen continue to probe and unlock the habits and migration of this mysterious fish.

The Rise and Fall of the Sardine Fishery

In 1925, the Pacific sardine (*Sardinops sagax*) was at the center of the largest fishery on the Pacific coast, with a catch of 173,000 tons (seventeen times that of the amount of albacore caught in the same year). By the 1936–37 season, sardines comprised the largest one-season landing of any single species ever recorded: 791,334 tons. But in less than a decade, the sardine population plummeted. They were gone. Fishermen lost their livelihoods, factories closed, and a thick malaise settled over the immigrant fishing communities.

How could one little fish bring such prosperity—at one time, it accounted for 25 percent of the total seafood catch in the United States—and yet be a major player in the almost total collapse of a once wildly prosperous fishery? A little biological background is in order. As one on the major forage species in the California Current, they have two characteristics that led to their exploitation: abundance and a tendency to form huge schools.

With regard to the former, it is not surprising that major sardine fisheries developed in California since seasonal upwelling planktonic blooms could support large populations of sardines. The schooling behavior of sardines allowed them to be easily caught, translating into relatively low operating cost for the fishermen and thus a relatively cheap product for the consumer.

The birth of the California sardine fishery began with the supply of whole fish in the 1860s, followed by the use of sardines as bait in the 1880s. These were modest undertakings, and the industry really took off with the shift to canning in the late 1880s. Sardine canning started in 1889 in San Francisco by the Golden Gate Packing Company. Even though it experienced some modest success, the plant inexplicably shuttered its doors in 1893. The remaining equipment was purchased by the Southern California Fish Company of San Diego, which canned sardines in oil, mustard, and tomato sauce in one-quarter-, one-, and two-pound tins.

But by early 1900s, Monterey would become the epicenter of the sardine fishery, with the startup of two plants: the Booth Plant in 1902 and the Monterey Fishing and Packing Company in 1906. The Booth cannery was built near Fisherman's Wharf, but Cannery Row would spring up farther out of town along Ocean View Avenue. The first major cannery on this boulevard was the Pacific Fish Company, founded in 1908.

Sardines (Richard Carey/Shuttestock.com)

With its growing infrastructure, it would not be long before Monterey sardines, harvested from the nearby fertile waters, would gain a reputation of having a flavor and quality equal to the then preferred French brand.

The coastal environs and surrounding enclaves would be settled mainly by Sicilian immigrants, who over generations had fished for sardines in their home waters. (The term "sardine" was first used in English during the fifteenth century and may come from the Mediterranean island of Sardinia, whose waters once supported an abundant population of sardines.) Sardine fishing was a family affair. Husbands could be found piloting the boats; their sons, as young as ten years old, would be aboard as mates; and their wives worked in the canneries. Working in the canneries was hard and tedious, involving long hours, sometimes twelve hours or more. The workday began with the arrival of the catch at night, and no one was allowed to leave until the entire catch was processed. Without any regulations, working conditions were often unsafe, and injuries were accepted as part of the job.

The boat of choice was known as a sardine launch, which was modeled after a salmon trolling boat. A typical launch was about thirty-four feet in length, was fashioned with a small wheel house, and was originally powered by a small gasoline engine, later to be replaced by a diesel engine. The stern was relatively low so the net could be retrieved easily.

In 1905, Sicilian fishermen introduced the use of the lampara net for catching sardines, and from 1906 for the next two decades, it was practically the only net used in sardine fishing in Monterey Bay. In principle, the lampara is similar to the purse net in that its bottom can be closed, but unlike the purse net, the lampara has a large "bag" or "sack" that holds the fish until the bottom can be closed. This bag, with its two long side wings, is laid out in a circle around the fish. As soon as the fishermen start to haul in the wings, the circle becomes distorted, and the net operates very much like a huge scoop. As the wings close, the fish are forced into the bag of the net.

To transport the catch to the canneries, lighters or barges were used on the fishing grounds. Lacking power of their own, they were towed by the launches. With a length between twenty-five to thirty-five feet and most of the deck space taken up with a large hatch, their capacity ranged from twelve to thirty tons of fish, with about twenty tons being the average load.

In 1922, the average lampara crew consisted of eight fishermen and the captain, who usually was the owner of the launch, lighter, net, and any incidental equipment. The captain was paid an agreed-upon price per ton of sardines delivered to the cannery. The money was generally divided into twelve shares for distribution among members of the crew, with each crew member receiving one share and the extra shares going to the captain to cover his investment and expenses.

The number of crews fishing for any one cannery depended primarily, but not entirely, on the capacity of the plant. While each crew, naturally, lobbied to have the highest sardine limit, from the canners' view it was more desirable to have more crews than was necessary and assign to each crew a smaller limit. The cannery owners reasoned that more boats fishing translated into a higher probability of the cannery meeting its goal. But there was no hard-and-fast rule for the number of crews; two canneries with the same capacity may have had different employment practices. As a result, the total number of crews fishing during a season varied widely. In the early 1920s, Monterey canneries may each have had only three crews fishing, but sometimes that number bloomed to fifteen or twenty crews.

As the demand for sardine products skyrocketed during the late

1920s, canners and fishermen responded accordingly. In September of 1924, there were seven operating canneries in Monterey, but by March of 1929, ten canneries dotted the shoreline. But even the older plants were not static, enlarging their floor space, adding new canning machines, and other equipment, which allowed them to greatly increase capacity. These changes in shore-based operations were accompanied by the employment of more fishing crews and increased fishing effort. The Monterey fishing industry was booming.

For Monterey, the peak sardine fishing was in the fall, and most of the sardine landings occurred within two miles of the Monterey canneries, particularly in the southern cove of Monterey Bay. In an in-depth study of the Monterey sardine fishery during the 1920s, W. L. Schofield includes an interesting map that depicts the locality of catches made by boats during the 1921–22 fishing season. There is a marked north-south gradient with regard to the concentration of catches in Monterey Bay: the northern reaches of the bay exhibit a dearth of fishing activity, probably due to the upwelling shadow effect, but the activity increases markedly to the south as upwelling becomes more prevalent.

Throughout the 1920s, the sardine fishery continued to grow, expanding southward to San Pedro and San Diego in California. By 1925, it was the largest fishery on the West Coast, landing 173,000 tons. In addition to producing a high-quality canned sardine for human consumption, the canneries expanded their operations into reduction, a process where canning waste was converted into protein-rich feed for chickens as well as fertilizer. So profitable was this undertaking that some plants started using the whole fish along with canning waste.

The California Department of Fish and Game and the U.S. Bureau of Fisheries became concerned about the direct use of sardines for non-human consumption and attempted through legal action to curtail the use of whole fish for reduction. The cannery owners were not willing to acquiesce, realizing that any limitation on the use of sardines would cut into their profits. They soon found that fish caught and processed beyond three miles were not subject to any of the state's laws. One Monterey canner was so bold as to tow a concrete barge outside the state's three-mile jurisdiction and commence reducing sardines, without even the hint of canning. Others would follow, and these ventures became so profitable that a whole fleet of floating reduction plants could be found anchored offshore from the canneries.

Such vessels operated with impunity well into the 1930s. In 1938, the state of California amended its constitution to stop offshore reduction plants. The effect was minimal since the reduction vessels had essentially stopped processing because of increased labor costs and a dwindling sardine population. In retrospect, the reduction ships were quite successful; in a nine-year period, they accounted for almost a fifth of all sardines landed during that period.

The 1936–37 season saw the entry of Oregon and Washington into the fishery. The shutdown of the reduction industry in California, however, provided an opportunity for the owners—some moved their processing equipment north to Oregon. In 1935, the Oregon legislature changed its regulations to allow sardines to be used for reduction. With this welcoming environment, four plants for receiving and reducing sardines to oil sprang up in Coos Bay almost overnight, followed by a fleet of seventy-five purse seiners from Monterey.

The landings of sardine in those states, along with those in California and British Columbia, yielded the largest one-season catch of any single species of fish ever caught in eastern Pacific waters—791,334 tons. As Randall Reinstedt wrote in *Where Have All the Sardines Gone?*, "There are enough 10-inch sardine in these landings that together, if laid end to end, would reach from the earth to the moon and back." The sardine population was under heavy fishing pressure.

A strong advocate for conservation measures was Frances Clark, who in the 1930s was the only woman in the United States to be in charge of a major fisheries research unit. In her 1939 paper, "The Sardine: International Aspects of Its Life History and Exploitation," Clark came to the conclusion that the sardines off California and British Columbia were the same stock. The fish that spawned in Southern California waters had by their second year migrated north. Clark lamented, "Throughout its entire life and along its entire range of distribution the sardine population is exploited by man." Her plea for proactive measures was a cry in the wilderness. She wondered how long the fishery would hold up.

From about 1934 to 1946, landings continued to be good, averaging almost 600,000 tons a season. Monterey continued to prosper, boasting nineteen canneries and twenty reduction plants in 1945. But biologists during this period were sounding the alarm that the sardine biomass could not sustain removals over 250,000 tons without a crash in the population. They argued for quotas, limits on catch. The warnings were ignored. World War II prompted good prices for sardines. As John Steinbeck writes in *Cannery Row*, "The canneries themselves fought the war by getting the limit taken off fish and catching them all. It was done for patriotic reasons."

The postwar period (1946–52) showed a marked drop in catch, a 40 percent decrease from the World War II years. The fishery, as biologists had predicted, was now in free fall. The 1952–53 season was the nadir for the California sardine fishery, yielding only 14,873 tons. The fishery, fueled by the eternal optimism of fishermen that the next season would be better, would struggle on for more than a decade. But the fish never returned in great numbers, and by 1968, the fishery was for all practical purposes a dying enterprise. Eighty years after its promising start in San Francisco, the fishery was gone. It had collapsed and subsequently disappeared over twenty-two seasons. The vessels were mothballed, the canneries closed their doors, and the people were no longer sardine fishermen, canners, or workers.

A few fortunate fishermen had the foresight, or maybe the luck, to see the handwriting on the wall during the late 1950s and early 1960s. They pursued other fisheries, such as Dungeness crab, rockfish, albacore tuna, and salmon. The vessels that entered into new fisheries would change from eight-man sardine crews to two- or three-man, or even captain-only, crews. The displaced crewmen had few employment options: either find work ashore, a hard pill to swallow for these life-long fishermen, or pull up stakes and head north to fish Alaskan waters.

Where does the blame lie for the collapse of the once wildly prosperous sardine fishery? While overfishing most assuredly was a contributing factor, historians are hesitant to make it the sole scapegoat, although the population probably would have rebounded more quickly had it not been so heavily harvested. In retrospect, management conflicts, weak legal policies, and climate change have to be given strong consideration. The synergistic effect of all these conditions most likely precipitated the sardine population crash.

Most fish populations undergo natural fluctuations in size. In an unexploited fishery, the size of the population is dependent on the carrying capacity—the maximum number of organisms that can be supported by the environment. Generally, the limiting factor in population growth is the availability of food. When food becomes scarce or the population is too large to be supported by the amount of food available, the result is an inevitable decrease in numbers, generally to be followed by a rebound in numbers as competition for a limited resource decreases.

Analysis of fish scale deposition in ocean basin sediments dating back 1,700 years indicates that sardine abundance fluctuates widely over a period of about sixty years. The scale deposition record shows nine major recoveries and subsequent collapses over 1,700 years, with the average time for recovery of the sardine population being thirty years.

While the scale deposition record is not sensitive enough to decipher yearly sardine biomass, it is possible that heavy fishing pressure coinciding with a down year for sardines may have led to a precipitous crash, preventing a normal recovery in numbers. Perhaps temporal or spatial changes in the California Current System initiated and enhanced the decrease in sardine numbers. Even today, we do not have the complete answer.

While the collapse of the sardine fishery sent shockwaves throughout California society, it did spawn the beginning of the systematic study of the California Current Ecosystem with the founding of the California Cooperative Fisheries Investigation (CaCOFI)—a collaboration of various research institutions, government agencies, and universities. CaCOFI's initial charge was to study the causes of the sardine decline, but since that humble beginning, the program has expanded to include many other species and to address other topics in the California Current—a decades-long force in attempting to unravel the mysteries of this ecosystem.

Are there any lessons to be drawn from such a historic collapse? John Radovich attempted to address this issue in the 1980s by analyzing the interactions of the politics of fishery management with the biology of the species. His sobering conclusion was that "the present scarcity of sardines off the coast of California, and their absence off the northwest, is an inescapable climax, given the characteristics and magnitude of the fishery and the behavior and life history of the species." An extension of his argument is that institutions, both governmental and commercial, adhere to the paradigm that "bigger is better," with size being the ultimate measure of "success," a formula that if followed leads to overfishing.

But greed and social and political pressures are not the sole culprits in this story. One word comes to mind: uncertainty. Uncertainty in the sense that resource managers are not able to determine with a high degree of confidence the fish biomass, the carrying capacity of the environment, how much fishing is too much, and whether an ongoing decline in a fishery is due to overharvesting or to natural causes.

An ancillary lesson to be drawn from the collapse of the sardine industry is that a substitute or alternative fishery will develop more rapidly than would be the case for a newly developed, independent fishery. The capital, labor, vessels, and technology of the sardine fishery were, for the most part, readily transferred to a substitute fishery. When technology and expertise are available, the "learning curve" for the substitute fishery is shortened or eliminated. The fishermen that moved from the sardine fishery to the alternative one were already skilled in fishing, familiar in the "ways" of the fish, and brought with them their cultural ties to fishing as a way of life.

What does the future hold for this iconic fish and a fishery that during its heyday, 1915 to 1951, landed 83 to 93 percent of the total catch along the Pacific coast? In 1999, fishery biologists cautiously declared the sardine resource fully recovered, with a spawning biomass estimated at over 1 million tons. (To assess sardine abundance, fishery managers are resorting to a traditional method—the use of small planes to find fish in the ocean.) Based on this positive data, a revived sardine fishery sprung up in 2000, but it in no way resembles its ancestors where a "Wild West" mentality permeated the fishery. Today, the fishery is highly regulated with limited entry and strict harvest quotas. Even the processing facilities now operate under stringent sanitary rules mandated by the federal government. The sardine canneries are all gone; the last sardine cannery in Monterey sold its canning equipment in 2004 and now processes only fresh and frozen products. The cost of doing business in California is high, and California product must compete in the market with imported fish produced at a much lower cost. Even with the changes, the sardine industry is still today a traditional fishery, but with a contemporary outlook.

Tackling the Gladiator of the Sea

The swordfish (*Xiphias gladius*) occupies its own branch (Xiphiidae family) on the billfish tree of life and is a single species worldwide, sharing many of the same morphological characteristics of its nearest relatives: a streamlined body, a long, pointed bill, and a pronounced crescent-shaped tail. But the dorsal fin of the swordfish is enough to distinguish it from other billfish—a high, rigid fin that is deeply concaved near its rear margin. And yet what catches one's attention are its eyes, black, tennis-ball-sized orbs stare back, mesmerizing you. As one old-time swordfisherman related, the eye "fixes you," rendering you useless to slay the beast.

Human fascination with swordfish dates well back into antiquity. Included in Aristotle's vast zoological research were swordfish, which he called *xiphias* (the sword). Polybius (203–120 B.C.), a Greek historian, describes, probably for the first time, in his *Histories* the hunt for swordfish. A highly migratory animal, the swordfish periodically entered the Mediterranean Sea and often could be found as far eastward as the Strait of Messina off the coast of Italy. As noted by Polybius, it was at this location where men would set out in small, flimsy boats to intercept the swordfish and harpoon it.

Swordfishing in California dates back more than 2,000 years with the Chumash, Native Americans who inhabited the Channel Islands. The advent of successful Chumash swordfishing hinged on two innovations: the plank boat and the harpoon. The Chumash's vessels were essentially twenty-foot driftwood canoes, which provided the speed and mobility to overtake a swordfish. The carved wooden harpoon bore a stone point on one end and a curved barb of deer bone behind.

To the Chumash, the swordfish was more than an animal to be hunted; it was an integral part of their culture. According to Chumash tradition, each creature of the sea had a land counterpart. The sardine, for example, was considered to be similar to a lizard. But swordfish were viewed in a much different light, as "people of the sea"—marine equivalents of human beings.

California's harpoon fishery began in the early 1900s. Prior to the 1920s, there was little consumer interest in swordfish, and landings were low. But by the mid-1920s, demand increased due to new markets developed in the Northeast. As consumers came to appreciate the swordfish's mild and sweet flavor, California landings increased from about 10 tons in 1925 to over 60 tons in 1927. For the next twenty years, landings fluctuated about a gradually increasing trend, peaking at approximately 622 tons in 1948. This "boom" year, as with the albacore fishery, was followed by a marked downturn in landings, when in 1950 the catch plummeted to only 16 tons. Some have attributed the decline to a shift in effort from swordfish to albacore. Similar to the albacore tuna fishery, the harpoon fishery in the Southern California Bight was mainly a seasonal venture, occurring during the summer and fall. Because this period overlapped with the arrival of albacore, some swordfishermen would sign up to fish for tuna if swordfish were scarce. But, of course, the simple truth of fewer fish to catch equals fewer fish caught cannot be ruled out to account for the precipitous drop in landings. Even with a banner year in 1978 (fishermen said the ocean was alive with swordfish, and landings peaked at 1,751 tons), over nearly six decades the Southern California harpoon fishery accounted for only 9 percent of the total eastern Pacific swordfish landings.

The Southern California swordfish fishery was modeled after the New England harpoon fishery, which had begun almost seventy years earlier. The harpoon, in particular, was almost identical in construction and appearance to that used by New Englanders. It was approximately fifteen to eighteen feet in length, fashioned out of hickory or some other hardwood. Often the bark was left on, so the harpooner might have a firmer handgrip. At one end of the harpoon, an iron rod (two feet) or "shank" was attached. Upon the end of the shank fits the head of the harpoon, a bronze dart that would be thrust into the side of the fish.

The early vessels were small sail-powered sloops or schooners. The harpooner was always positioned at the end of the bowsprit of the vessel, nestled into a raised platform or "pulpit." As an aid in locating swordfish, a second platform, the crow's nest, was located high on the vessel's mast. Manned by a crew member, equipped with the keen eyes that practice had given him, he was capable of sighting a swordfish more than two or three miles in the distance. Upon a sighting, the watch, from his precarious position, sang out, alerting the captain to steer the vessel toward it.

The vessels that participate in today's swordfishery, while bigger, refrigerated, and diesel powered, maintain the basic components of their predecessors. And the harpoon fishing gear has changed little since the 1900s except for minor modifications in the metal used in the shank.

Once a fish is sighted, the ship is maneuvered so that the plank—a twenty- to thirty-foot-long scaffolding extending from the bow of the vessel—is over the fish, and from his pulpit at the end of the plank, the harpooner strikes the fish. The handle is pulled free from the dart, and the mainline, marker flag, and floats are thrown overboard. The fish is left to die, and after an hour or two, it will be retrieved to be dressed out and stored on ice. During the early years of the harpoon fishery, the livers were sometimes removed and sold because of their high content of vitamins A and D—a practice no longer followed because of modern pharmacology.

In the early 1970s, the commercial swordfishery started to employ

Swordfish harpoon boat (© Istockphoto.com/Gaspare Messina)

the use of airplanes, which not only expanded the search area but also made it easier to spot the fish. Records for the 1974–75 season indicate that aircraft-assisted vessels increased their landings threefold from the previous year. In 1974, the California Fish and Game Commission, under pressure from a coalition of sportfishermen and more traditional harpooners, prohibited the use of aircraft effective June 28, 1976. But in November of 1976, the restriction was relaxed to allow aircraft to be used only for reconnaissance and not as an aid to the harpooner.

The success of the harpoon boats depended on one critical factor: the tendency for a swordfish to bask near the surface. With its high dorsal and tail fins clearly visible, the swordfish became an easy target. The swordfish is the only known billfish to exhibit this habit, and it appears to be related to the fish's digestion and internal temperature.

When not at the surface, swordfish can be found probing the cold depths for food. To help speed up digestion, the rates of which are a function of temperature, the swordfish may opt to bask in the warm surface water. (For example, a swordfish in 42°F water has a digestion rate that is approximately three times slower than in 68°F water.)

Though the swordfish is known to conserve some body heat, its core body temperature, over time, decreases during prolonged dives. The swordfish must come up from the cold to warm itself.

A New Player Enters the Game

By 1980, the use of drift gill nets had entered into the California swordfishery. Until 1982, the gill net fleet was limited to an area south of Point Conception, but by the summer of that year, the fleet had expanded its operations northward to Morro Bay and increased in size to over 200 vessels. The result of this massive effort was predictable: swordfish landings by the gill net fleet were a record 2,400 tons in 1985. And just as predictable was its effect on the harpoon swordfishery: harpooning on the California coast became a fishery in decline. Harpooners could not compete with drift net fishermen in the numbers of fish caught. And yet a small number of harpooners have hung on; even today, they still pursue swordfish much the same way their predecessors did.

While harpooners grudgingly acknowledge that drift nets are *effective* in catching swordfish, they vehemently disagree with drift net proponents who extoll the *efficiency* of the nets. To harpooners, efficiency means getting the job done with minimal damage or disruption. The nonselectivity of gill nets in catching myriad species is a rallying point for harpooners. They have vehemently argued that harpooning is both a sustainable and selective way of catching fish. With a harpoon, the fishermen can target only mature swordfish, bypassing undersized, juvenile fish. A group of California harpooners have also attempted to create a niche market for harpooned swordfish, promoting a high-quality, fresher product to a more discerning consumer.

But while financial gain may be the prime motivating factor for some to plug along in what many consider an archaic venture, Zane Grey may have captured the essence of swordfishing in his short story "Swordfish," which was ultimately published in *Tales of Fishes*. "The pursuit of the swordfish is much more exciting than ordinary fishing, for it resembles the hunting of large animals upon land and partakes of more of the chase. . . . The game is seen and followed, and outwitted by wary tactics, and killed by strength of arm and skill." An old swordfisherman who had plied his trade for over twenty years related to Grey that while sleeping, "many a time he had rubbed the skin off his knuckles by striking them against the ceiling of his bunk when he raised his arms to thrust the harpoon into visionary monster swordfishes."

But in many cases, the competition from gill nets proved too much for some harpooners, and many converted their vessels to drift gill net gear or obtained permission to use both types of gear. In 1979, 310 permits were issued to harpoon swordfishery, but by 2001 that had declined to only twenty-five.

Today, the majority of swordfish caught in California waters are by means of gill nets. Due to the negative publicity that the drift net fishery has come under since its inception, it is now a strictly regulated industry, including area and time closures, with onboard observers to monitor catch and operations.

Another Blow to the Swordfish

In 1991, three longliners from the Gulf of Mexico arrived in Southern California to assess the potential of fishing for tuna and swordfish ostensibly outside the 200-mile limit of the U.S. Exclusive Economic Zone. In 1991 and 1992, respectively, these vessels landed 28 and 29 tons of swordfish. By August 1993, other vessels from the Gulf of Mexico began arriving in Southern California, and consequently landings jumped to 101 tons. By 1994, the number of longliners increased to thirty-one, and reported landings topped 500 tons. The concern for many fishery managers was that these vessels were now heavily impacting the eastern Pacific stocks that are integral to the California inshore fishery, particularly that within the Southern California Bight.

With expansion of this fishery and indications that its growth will continue, accurate stock assessments are critically needed. As of now, these assessments have a large measure of uncertainty that needs to be resolved if fishery management options are to be proposed, evaluated, and implemented.

Squid Fishery

While probably not having the charisma of the powerful swordfish or the speedy albacore tuna, the California market squid (*Loligo opalescens*)—a small (twelve inches) marine invertebrate—is one of the state's most commercially valuable species, generating millions of dollars of income from domestic and foreign sales.

Loligo opalescens is one of thirty to forty species of squid in the Loginidae family, but in California, it is the only squid species consistently taken for commercial purposes. The majority of the catch is frozen and exported to China, Japan, and Europe for human consumption. Domestically, market squid is utilized as either a canned or a fresh product as well as bait for recreational anglers.

Though market squid range from Baja California to Alaskan waters, they are rarely available in fishable numbers north of Vancouver Island. Their major spawning site appears to be in the nearshore waters off Central and Southern California, where large spawning aggregates can be found from November through April. Due to their short life span (twelve to eighteen months), market squid reproduce at a young age, making then an ideal fishery since they are resilient to fishing pressure.

As with sardines, Monterey Bay became a major squid fishing locale, with the Chinese the first to harvest them in the late 1880s. Setting out at night in small skiffs, called sampans, into the shallow waters of the bay, Chinese fishermen lit torches or hung wire baskets with burning pitchwood over the sides to attract these nocturnal feeders to the surface, where nets could be set around the squid schools. The fishermen dried the catch, after which it was mostly shipped to China, but some of the squid found its way to markets in San Francisco.

In 1905, Italian immigrants introduced their lampara nets to the fishery. To lift the squid out of the main net, around 200 pounds at a time, into the boat, fishermen employed a smaller brail net that was used well into the 1970s.

While the competition between the Italians and the Chinese for squid ultimately forced some within the latter group to seek employment in the processing and export sectors of the fishery, it led to the growth of market squid as a major product from Monterey Bay. On the heels of the Monterey fishery, a Southern California one developed, as Italians and Yugoslavians settled the Santa Barbara and San Pedro areas. For the most part, these fishermen eschewed the use of the lampara net in favor of the brail net to bring the catch onboard; no other nets were used. Vessels employing this method were known as "scoop" or "brail" boats, which tended to be smaller than the lampara or purse seine vessels. Over time, fishermen realized that the smaller brail boats could not compete with the larger seine vessels, which could easily meet increasing market demands for squid. Around 1977, the southern fleet made the shift from brail to purse seine.

In today's fisheries, the vessels include a light boat, which uses high-wattage bulbs to attract and concentrate the squid near the surface, and a seining boat, which deploys a net to encircle the aggregation of squid. In place of the antiquated brail net to scoop the fish into the boat's hold, seiners extract the fish from the net by means of suction generated by a centrifugal pump. From 1996 to the present, approximately 95 percent of the vessels have used some type of seine net, and only 5 percent used brail nets.

While Central California squid fisheries provided the lion's share of the catch to markets before World War II, by 1960 and up to the early 1980s, landings were equally divided between Central and Southern California. But a seismic shift occurred in the late 1980s when the Southern California fishery easily outpaced its northern counterpart in annual catch. Since 1985, the Southern California fishery has dominated statewide landings and expanded its fishing range, particularly in the Channel Islands. By 1993, market squid was the largest commercial fishery by volume (47,100 tons), and by three years later, it had become the most valued fishery resource, generating some $33 million.

Coinciding with the substantial growth in the California squid fishery—landings increased almost 400 percent from the 1990–91 season to the 1997–98 season—managers expressed concerns regarding its overall ecological and socioeconomic sustainability. Prior to 1997, regulations were piecemeal and limited to the Monterey Bay fishery, where, for example, the use of lights as an attractant has been permitted and banned many times since the inception of the fishery. Fishermen who had lost their jobs in other fisheries because of declining stocks flocked to the growing squid industry, one of last remaining open-access fisheries along the Pacific coast. While they were met with open arms by suppliers, who needed more fishermen and vessels to meet the demand for squid, these newcomers faced opposition from local fishermen who felt that their livelihoods might be threatened given overexpansion of the fishery. The locals were especially hostile to those fishermen who were from other states and perceived as having no stake in the long-term vitality of the squid stock.

The rapid increase in harvest and number of new vessels entering the fishery prompted some within the industry, particularly Monterey fishermen who favored a limited-entry fishery, to seek legislative recourse. Beginning on April 1, 1998, a moratorium was placed on the number of vessels participating in the fishery, and an annual permit fee was implemented for three years in order to fund assessments of resource abundance that could guide managers in setting conservation measures. Under these new guidelines, the first fishing season (1998–99) had 243 purse seiners and fifty-three light boats in the fishery.

By 1999, another concern about the fishery was raised by the National Park Service (NPS) with regard to the effect that light vessels were having on seabirds nesting on the Channel Islands. The NPS produced evidence that excessive lighting was resulting in increased seabird nest abandonment and chick predation. The California Fish and Game Commission (CFGC) studied the evidence and imposed statewide wattage restrictions and shielding requirements on the squid fleet. Even with these restraints, the fishery experienced a banner year in 2000 with record-high landings.

In 2004, the CFGC implemented the Market Squid Fishery Management Plan (MSFMP) to ensure long-term sustainability of the resource and to provide a management framework that would be responsive to socioeconomic changes in the fishery. The plan had a wide-reaching agenda: to set a seasonal catch limit; to develop a restricted access program based on historical participation in the fishery; to restrict the use of lights near seabird nesting sites in the Farallon Islands; and to establish an advisory committee composed of scientific, environmental, and fishery representatives.

So what is the state of the fishery now? California has a major fishery for a species that appears to have unlimited market demand. But since the implementation of the MSFMP, questions remain, many of them holdovers from the 1970s. What is the fecundity of squid? What is the survival rate of eggs released into the environment? How large is the larvae population? The answers to these questions are critical because the market squid fishery takes place on the spawning grounds, and it is imperative that all interested parties allow enough eggs to be spawned before harvest. This conservation measure will ensure a sustainable resource well into the future.

The Push to the North

While California's fertile waters proved to be an irresistible lure for many Europeans and Asians, some hardy entrepreneurs sought their fortune in the Pacific Northwest. Their quest was not for gold or timber or furs but for salmon.

Although Native Americans had fished the salmon runs from Northern California to Alaska for thousands of years, their impact on the salmon population was minimal. Theirs was, as was that of many Indian tribes of the Pacific coast, a sustainable harvest. Over time, they developed and implemented a set of guidelines and customs that functioned to limit their catch—the first attempt at salmon management in the Pacific Northwest.

But the influx of nonnative fishermen to the Pacific Northwest would mark a sea change in how salmon were caught and managed. The local tribes used a variety of fishing gear, including spears, weirs, and traps, to catch salmon migrating up the river. In contrast, nearly all the early fishermen employed heavy linen gill nets, which stretched hundreds of feet across the river. The nets proved very effective in ensnaring salmon since they were deployed at night, when the meshes of the net were essentially invisible to the salmon. The fishermen boasted how easy it was to catch the salmon—a brag they would ultimately regret.

As more fishermen settled into the region and in the absence of effective government oversight, the salmon fishery evolved into a freewheeling, unregulated, and highly competitive endeavor. Ethnic communities fought one another for prime fishing sites where they could set their nets. A free-for-all mentality set in, sometimes with drastic consequences. Chinese fishermen, in particular, were prevented from entering the salmon fishery. Those who attempted to enter the fishery often found their boats scuttled, their nets shredded, and their lives endangered. Even the indigenous peoples were treated as outcasts, displaced from their traditional fishing haunts, and blamed for the decline of salmon catches.

The early salmon fishery was confined to the many rivers that drained the land, but in 1898, F. J. Larkin had an idea that would lead to a seismic shift in the fishery. His intent was to replace the picturesque butterfly sails that powered the gill net boats in the Columbia River with gasoline engines. Over the next few years, the use of gasoline engines in the fishery spread like wildfire. The engines not only increased the effectiveness of the gillnetters but gave rise to a whole new method to catch salmon—trolling. Trolling vessels became lethal hunting machines, capable of deploying over 100 lines with a lure or bait to attract hungry salmon. But probably their main advantage was mobility; they were now able to motor over large swaths of water. By 1920, estimates place the number of trollers in the sea off the Columbia River at approximately 2,000 vessels.

Salmon could be caught in both rivers and the sea because they are anadromous. Salmon are born in freshwater, spend most of their lives in the open ocean, and return to freshwater to spawn. But what advantage did the salmon achieve by migrating between freshwater and saltwater during their lifetime? Surely, the advantage had to outweigh the accrued physiological and behavioral costs. A young, stream-dwelling salmon is well adapted to its environment, sporting natural camouflage that affords it a degree of protection from predators. In seagoing salmon, this coloration gives way to uniform silver on the sides and the underbelly—one better suited to the monochromatic world of the open sea. Salmon must spend some time in freshwater to develop the capacity to osmoregulate—maintain an internal balance between water and dissolved solids, such as salts—before entering the ocean. For pink salmon (*Oncorhynchus gorbuscha*), their time in freshwater is very short as they migrate to the sea as free-swimming juveniles. Sockeye (*Oncorhynchus nerka*), coho (*Oncorhynchus kisutch*), and some Chinook (*Oncorhynchus tshawytscha*) salmon remain in freshwater for up to two or more years before the urge to swim downstream overcomes them. In spite of all the stresses, the migration across the fresh-saltwater boundary is driven by one overriding factor: the availability of food. The northern waters of the Pacific are more productive than the adjacent freshwaters.

As more and more trollers took to the open ocean to catch salmon, concerns deepened among conservationists and government officials about the impact of this essentially unregulated fishery on the health of the salmon population. As early as 1923, biologists came to the sobering conclusion that salmon could not survive both the river and ocean fisheries. In particular, the ocean fishery was cited for catching immature fish before they reached their full size. Chinook salmon, the largest of the five species of Pacific salmon, routinely reach weights of twenty to thirty pounds as adults. But trollers that headed out to the feeding grounds of the salmon were hooking salmon in the range of five to seven pounds. The more the practices of the trolling fishery were probed, the more "red flags" were raised. In particular, opponents of the fishery questioned exactly whose fish the trollers were catching. To answer that question, researchers needed to determine the prevalent salmon migration patterns, where in the ocean salmon from specific rivers spend most of their adult lives, and who was catching them. Tagging studies conducted by the Canadians between 1925 and 1930 confirmed that salmon migrated long distances—some far out into the Pacific subpolar gyre—and that trollers intercepted fish from different rivers. Armed with this evidence, various constituencies voiced opposition to trolling. In response, Washington and Oregon banned trolling within three miles of the coast but were powerless to outlaw the practice outside the state jurisdiction of three miles or to regulate out-of-state fishermen. Without any legal restrictions, the troll fishery continued to grow, gained political clout to foster its goals, and all the while plundered the resource. It was not until 1949, when California, Oregon, and Washington brought the Pacific States Marine Fisheries Commission into existence, that some degree of control was exerted on the fishery.

While some problems were solved, others arose. The advent of pelagic netting in the Pacific by Japanese and Taiwanese fleets led to the indiscriminate capture and killing of salmon. To effect some regulation and establish boundaries for open-ocean fishing, in 1952 Congress adopted the International Convention for the High Seas Fisheries of the North Pacific. An agreement, known as the Trilateral Pacific Salmon Treaty, between Japan, the United States, and Canada, set forth the "abstention principle," which allows the nation where anadromous fish originate to prohibit fishing of these stocks on the high seas. Japan agreed to stop fishing Canadian or Unites States salmon east of the 175th meridian, roughly in the center of the Pacific. Over time, the treaty has been renegotiated and modified (the line shifted 10° of longitude to the west).

The salmon is such a highly prized fish that it was inevitable that other conflicts would arise. During the 1970s, the Alaskan troll fishery was, and still is, primarily an intercept fishery, with the majority of its Chinook catch originating from salmon runs in British Columbia, Washington, and Oregon. Alaska's refusal to reduce its harvest to meet Canadian management goals made political conflict almost inevitable. The Canadians viewed Alaska's inflexible position as a time bomb waiting to explode, one that would endanger future salmon stocks. As a counterargument, U.S. officials argued that Canada's demand for a reduction in the Alaskan catch was weakened by a Canadian desire for a larger share of the harvest. Some headway was made in resolving this thorny issue in 1985 when Canada and the United States signed the Pacific Salmon Treaty, which would regulate and allocate harvests within the Chinook salmon fishery while conserving stocks in British Columbia. Though viewed as landmark legislation to resolve the fishery wars, the treaty did not prevent future disputes, which still threaten the salmon—a species that now must confront many other obstacles.

A Sportsman's Quest

It is not a far stretch to state the history of big-game fishing began with establishment of the Tuna Club of Avalon on Catalina Island. It is in the waters off Catalina Island where the world's first rod-and-reel capture of tuna, marlin, and swordfish took place and caught the attention of a nation hungry for adventurous fishing exploits. All the essential ingredients came into place to make Catalina Island the epicenter of the pursuit of big, strong, and tenacious fish: abundance of large fish that in some cases reached weights of over 1,000 pounds, the evolution of fishing tackle, and the arrival of men willing to pit their skills and determination against many of the apex predators of the California Current.

One of these men was Charles Holder, who would often venture out to sea to tackle pelagic behemoths. After suffering a few disappointing seasons during which he was unable to catch tuna, partly as a result of his inexperience to catch such fish, in 1898 Holder successfully boated a 183-pound tuna, which was hailed by the small cadre of sportfishermen as the first "very large one." News of his accomplishment spread quickly, and for the first time, the public took a keen interest in big-game fishing.

Buoyed by his angling feat, Holder, with the support of his colleagues, founded the Tuna Club in 1898. With an emphasis on ethics and sportsmanship, the club's motto of "Fair Play for Game Fishes" was unanimously endorsed by club members. Any infraction of the club's rules could, and did, lead to anglers losing their club membership.

Holder's commitment to ethical angling standards may have had its roots in the thinking and writings of Izaak Walton. In his seminal work, *The Compleat Angler* (1653), Walton strongly promotes the sport of fishing from a social and moral aspect. Holder, as a transplanted easterner who learned to fish the creeks and rivers of New England for trout, was most likely familiar with Walton's work.

The strict code of angling conduct was not a deterrent to attracting a wide variety of sportsmen, dignitaries, and personalities from becoming members of the Tuna Club. From the political arena, Theodore Roosevelt and Winston Churchill were on the guest list, and Hollywood entertainers, including Charlie Chaplin, Jackie Coogan, Bing Crosby, and Stan Laurel, gravitated to the club. In 1936, Laurel showed that his fishing prowess was just as impressive as his comedic genius by catching a 258-pound swordfish.

While Laurel's catch was impressive at its time, it was not the first swordfish to succumb to an angler. That distinction goes to William Boschen, who, twelve years before Laurel's catch, caught a swordfish on rod and reel off Catalina Island. He was able to land the 358-pound swordfish in about ninety minutes, a testament to his skill and strength. Over time, Boschen would be recognized by his peers as a master angler, defining the art and the herculean effort needed to subdue a swordfish in a sporting manner. Zane Grey, who was an accomplished angler in his own right and fellow club member, viewed Boschen as the model that all anglers should attempt to emulate, one who could catch a swordfish in a fair battle. To Grey, swordfishing was the ultimate test for the angler, requiring more skill, nerve, endurance, and strength than hunting a grizzly bear or a jaguar, both of which he had killed.

Grey was not alone in his respect for swordfish. Human fascination with this species dates well back into antiquity. An ancient legend in Greek mythology tells of how the Myrmidons, followers of Achilles, took part in the Trojan War, in which Paris slays Achilles. Beset by rage and despair for their fallen leader, the Myrmidons hurl themselves into the sea. Thetis, Achilles's mother and a sea goddess, transforms them into fish. But in recognition of their sacrifice, she allows them to keep their swords but changes them into long bills on the snout, giving rise to the swordfish.

During his time on Catalina Island, Grey focused solely on swordfish. On his boat, he kept only one rod and reel, which was rigged exclusively for swordfish. To aid in his ability to see surface-cruising or basking swordfish, Grey installed an observation platform about two-thirds of the way up the mast. But despite his relentless efforts, he achieved only modest success, suffering many frustrations. During one season, he spent ninety-three days fishing off Catalina, during which he was able to catch only 4 of the 140 fish he sighted.

In spite of his less-than-stellar record in landing swordfish, Grey remained an avid proponent of swordfishing, spreading, whenever possible and to whoever would listen, the gospel of big-game fishing far and wide. When his brother, R. C., caught his first swordfish, some 1,000 spectators turned out for the weighing of the fish back at the dock.

The killing of a swordfish, or for that matter any large pelagic fish, and unceremoniously hanging it by its tail to be gawked at was a common practice during Grey's time, and one that continues in some circles even today. Some have argued that the killing of one of God's creatures is an abomination, an act perpetrated by egocentric anglers. To argue that fishing is a sport or that the fish does not suffer is to obscure the fact that we are taking its life. But history offers a different view of the complex relationship between humans and animals. Some historians have argued that this act of killing another species that will not be eaten or provide some other utility is rooted in early Western philosophical tradition that denies any moral relationship exists between humans and nature. The two philosophers most associated with this tenet, Aristotle and Thomas Aquinas, proposed views that demonstrated minimum interest in assigning moral status to any forms of life except humans. Aristotle espouses the following in his *Politics*: "All other animals exist for the sake of man, for use he can make of them." Sixteen centuries later, Thomas Aquinas picks up this theme but couches it in a more theological context: "We refute the error of those who claim that it is a sin to kill brute animals. For animals are ordered to man in the natural course of things, according to divine providence. Consequently, man uses them without any injustice, either killing them or employing them in any other way."

Another take on the killing of another being is expressed by the early Native Americans of the Pacific Northwest. According to their beliefs, when a hunter killed a salmon or deer or bear, the death was not a sign of human superiority. Rather, the killing was viewed as a gift of food, readily given by the animal to man. The animal allowed the human to slay it, and the hunter assumed the responsibility to treat the animal (gift) with respect. Fish, as were terrestrial species, are thus seen as a commodity to be fully used by humans. They are not perceived as having any identity as living organisms.

There is no record of Grey's personal opinion about the killing of fish, but that does not mean that Grey was totally callous toward the creatures he pursued. One event stands out. On July 1, 1919, Grey's brother hooked a swordfish estimated to weigh over 700 pounds. After a prolonged battle of more than eleven hours, the line parted. Grey was appalled to think of such a large fish swimming off, if it indeed survived, with a hook embedded in its mouth and trailing many tens or hundreds of yards of line behind it, almost certainly consigning it to a slow and painful death. From this experience, Grey would vehemently argue that the use of fishing tackle that is too light, as was the case above, is a sin: if the fish does not break the line, it must be fought for hours on end until it dies of exhaustion. His core belief, which was in line with the principles of the Tuna Club, was to fight the fish in a sporting manner, permitting the fish to jump and run as freely as possible. Without maybe realizing it, Grey was possibly showing respect for the fish that he attempted to catch but which often eluded him.

The degree of fight in a fish, as Grey most assuredly pondered, is now known to depend on the functioning of its muscles, which, in turn, depends on an energy source and oxygen. What leads to a fish succumbing to the stress of being hooked are the physiological limitations of how much energy and oxygen the heart and blood can transport to the muscles. As the fish struggles against its adversaries, the oxygen in the muscles becomes depleted, the energy supply is exhausted, and metabolic wastes, particularly lactic acid, accumulate in the muscles. If the fish, one already stressed due to the encounter, is unable to eliminate lactic acid, extreme muscle fatigue can result, and ultimately the fish could succumb to its weakened state.

From Grey's experiences and those of other members of the Tuna Club would come a revolution in fishing tackle, which would not only increase the odds of an angler landing a big fish but also allow the angler to undertake the battle in a sportsmanlike manner. William Boschen is credited with the original concept for the internal star drag reel, which allowed the angler to exert significant pressure on the fish, thus decreasing the duration of the fight. A prototype was made in Brooklyn, New York, by reel manufacturer Julius vom Hofe and was used by Boschen to catch that first swordfish.

In 1911, Captain George Farnsworth, who piloted the boat on many of Boschen's fishing outings, introduced the practice of suspending bait from a kite, which would increase the hookup rate of finicky fish. Previously, the boat's noise and shadow had spooked skittish fish, leading to few catches. Or possibly the fish had learned to recognize the boat as a dangerous predator. Kite fishing enabled a boatman to steer clear of a school of fish while keeping the angler's bait skipping enticingly over the water surface.

By the 1920s, Grey was routinely using a prototype Coxe reel, especially made for him at the then exorbitant cost of $1,500, which was much sturdier and larger than those in current use, allowing Grey to pack on more line. To complement the reel, he also had special lines, stronger and longer, made for him. All this gear was to be used for his single pursuit—swordfish. In 1926, Grey proved his instincts were correct, catching a 582-pound swordfish with the heavier tackle.

From Holder to Boschen to Grey, these men would provide the foundation and impetus for a major sportfishing enterprise in the twentieth century. In 1950, two brothers from Newport Beach changed the nature of local fishing by using barges and other boats for fishing. They became known as "party boats" or "cattle boats" because they could accommodate a large number of anglers. Part of the appeal of Newport Beach is the presence of a submarine canyon, right off the coast, which is a mecca for game fish. While anglers, many of them veterans of World War II, were happy to catch a variety of species, including black fish and yellowtail, others dreamed of more exotic locations where subtropical and tropical fish, not normally present in Southern California, were big and plentiful.

One of these dreamers was Bill Poole, who was a pioneer of the long-range fishing trips that originated in San Diego. In 1951, Poole piloted the fishing vessel *Polaris* out of Fisherman's Landing to begin a seven-day journey to Guadalupe Island, located approximately 220 miles south of San Diego. On board were thirteen adventurous anglers who had each paid $175 to fish this remote outpost. Guadalupe Island is an extinct volcanic island surrounded by deep water, with depths approaching over 12,000 feet.

Guadalupe Island, as well as similar features along this stretch of the Mexican coast, can change the physical circulation and/or the biogeochemical cycles resulting in greater food supplies, which attract apex predators. And so to Poole, Guadalupe Island was not a haphazard choice but one that offered the potential for outstanding fishing. While amenities aboard the *Polaris* were limited—no refrigeration, all ice; no showers, saltwater wash-downs—the superb fishing proved to be worth all the inconveniences of the arduous trip.

The vision and enthusiasm of these early advocates of long-distance fishing would spur on others to enter into the game. But competition and curiosity led to the search for new and better fishing grounds. In 1970, Alijos Rocks—a small group of tiny and steep pinnacles of volcanic rock located 500 miles south of San Diego—became the new "hot spot." On first view, they are unimposing—desolate rocks jutting only 110 feet above the water. But you are only seeing the summit of an undersea mountain that is close to 12,000 feet high. At this site, interactions between wind and tidal currents and topography drive nutrients to the surface and sustain high biological productivity. These barren rocks, in turn, become a haven for marine life.

Throughout the 1970s, the push southward would continue, and it seemed that the ultimate tuna destination had been discovered—an island chain 1,000 miles south of San Diego known as the Revilla-gigedos. The four islands and a submerged seamount known as Hurri-cane Bank seemed to offer unlimited possibilities. And it would not be long before those optimistic outlooks proved to be correct. With histori-cally very little or no fishing pressure to speak of, the islands yielded a cornucopia of fish, including hard-fighting wahoo, dorado, and the occa-sional marlin. But the main prize was giant yellowfin tuna. In 1979, on a seventeen-day trip, a 388-pound world record fish was taken aboard the *Royal Polaris*—a bigger, better-equipped version of the original *Polaris*.

For years, the yellowfin fishing in the Revillagigedos had no equal. As Bill Poole fondly recalled, "If we came back half-full of fish, seven or eight tons (caught by about 20 passengers), it would be a horrible trip. On one trip to Clarion (the southernmost island in the chain), we filled the hold in three days. We were filleting and putting them in the freezer as we ate up the food supplies. After a while, we wouldn't keep anything under 100 pounds. Then it was 200 pounds. Believe me it wasn't easy to tell a guy with a 150-pound fish that he had to let it go."

But the halcyon days would not last forever. Drawn by the abundance and size of the yellowfin found in the waters surrounding these islands, commercial fishermen, including purse seiners and longliners, entered into the fishery. The results were not surprising: unsustainable catches of tuna as well as unacceptable levels of bycatch of resident fish, such as grouper and pargo. Alarmed by the impact on this vibrant ecosys-tem, the Mexican government closed the islands to all fishing in March 2002, establishing a protected biosphere. The San Diego–based long-range fishing fleet was thrown into an economic tailspin. After years of negotiations with Mexican authorities, some progress was made to re-solving this thorny issue when Captains Tim Ekstrom and Randy Tous-saint, owners and operators of the *Royal Star* out of Fisherman's Land-ing, were granted fishing permits in 2006 and 2007 for catch, tag, and release fishing of yellowfin tuna. As pointed out by Ekstrom, this would be the first time that anglers would be paying strictly for the sole pur-pose of tag-and-release fishing. And the anglers would pay dearly to par-ticipate in an essentially scientific project, working together with per-sonnel from the Inter-American Tropical Tuna Commission and the Instituto Nacional de la Pesca. For this seventeen-day oceanic safari, each angler would shell out $4,000 for the opportunity to catch as many fish as they could for 100 percent tag and release and another $1,500 to pay for eighteen retrievable electronic archival tags that would be surgically implanted in select yellowfin. (Archival tags, which record depth, light, ambient temperature, and the fish's internal temperature at two-minute intervals and store all the data, were first tested by noted tuna researcher Barbara Block and her co-researchers, who proved they could haul a live bluefin tuna from the waters off North Carolina onto the deck of the boat, implant an archival tag in its gut, and release it alive.)

In spite of the cost, a far cry from the $175 charged by Poole for the initial long-distance trip, eighteen anglers signed up for this opportunity to haul back on one of the strongest fish in the ocean; one was Len Cunningham, who had fished with Poole back in the 1950s. Cunningham is an old-school angler, fishing during a time when no fish were released.

Though Michigan fly anglers instituted the practice of catch-and-release fishing during the 1950s as a means of reducing the practice and expense of stocking trout in a stream, it was slow to gain acceptance in the saltwater angling community. The practice butted up against the widely held belief that the oceans had an unlimited bounty of life—that it would be impossible to deplete these resources. But Cunningham was onboard this excursion because, though he remembers a time when "we had all the fish we wanted to catch and there were no limit," he had arrived at the stage in his life where he enjoyed catching and releasing big tuna.

Captain Frank LoPreste, who owns Fisherman's Landing along with Poole and operates the long-range boat *Royal Polaris*, supported the trip but had his reservations: would it prove to the Mexican officials that sportfishing has minimal impact on the sustainability of pelagic fish in the Revillagigedo archipelago? Would older anglers stay motivated to catch and release so many fish—about twenty-eight yellowfin and seventeen wahoo each—to reach the scientific goals of the expedition without being able to keep a single fish?

LoPreste was in favor of any research trip keeping a few fish. While Ekstrom agreed, a restricted research trip, for now, was the only way to access the islands. He believed that "this is our opportunity to fish the Revillagigedo Islands alone, without competition and to break new ground and begin a whole new relationship with the Mexican government and their scientists."

The gambit would pay off. By 2010, the islands were opened to fishing, although on a permit basis. But even before the islands were closed off to fishing, some were pushing the boundaries of long-distance fishing even farther.

Frank LoPreste made a twenty-three day trip to Clipperton Island, a tiny, French-owned atoll located over 1,600 miles south/southeast of San Diego. Many considered this trek way too risky, but LoPreste returned to San Diego with the twenty-seven-ton hold of the *Royal Polaris* filled with fish. It was the biggest haul in the history of Southern California sportfishing.

Over the decades, the long-range fleet has grown not only in the number of vessels but also in the size of the vessel (up to 124 feet long and 32 feet wide), and competition for anglers has forced owners and captains to find ways to improve the quality of the trips. Amenities now include private staterooms, air conditioning, gourmet meals, plasma-screen TVs, multiton fish holds, and state-of-the-art refrigeration. These vessels have become floating sportfishing laboratories, experimenting with new tackle and refining angling techniques, which most assuredly would have appealed to Zane Grey and the original members of the Tuna Club.

While anglers continue to seek out new fishing grounds in pursuit of ever bigger fish, a potential problem looms on the horizon: vast swaths of the eastern tropical oceans are experiencing marked decreases in dissolved oxygen concentrations. The concern is that highly mobile game fish, such as swordfish, marlin, and tuna, require large amounts of dissolved oxygen to adequately meet their metabolic and physiological demands.

While oxygen-minimum zones are naturally occurring hypoxic layers found at depths between 600 and 3,000 feet and generally are the lower boundaries for pelagic fish movement, these hypoxic waters have expanded both horizontally and vertically, decreasing, for example, the habitat of Pacific billfish by 15 percent. Because oxygen deficit can be a major detriment to a robust fishery, the rise and expansion of hypoxia represents a major perturbation to the structure and functioning of marine ecosystems.

Chapter Five

SEA OTTERS & PINNIPEDS

From my vantage point on the vessel, I am first to spot the sea otter, as it bobs up and down with each passing wave. This sea otter, a member of the weasel family, is contently lying on its back amid the thick fronds of kelp that surround it. On its stomach lies an assortment of objects it has gathered from the ocean floor, including an abalone and a hefty stone. Lifting the abalone with its forepaws, which when I look at them appear to be too stubby to be of much use, the otter deftly hammers the hard-shelled mollusk on the rock. After a few blows, the shell gives way, and the otter scoops out the fleshy innards. The otter will continue to smash abalones and eat until satiated, but that could take a while. A sixty-pound otter will eat 25 percent of its body weight every day. Once full, the otter will nap. Its blanket is the kelp, which it rolls around its body, anchoring itself in place in the rolling sea. But it would not be long before our sea otter has some company. Of all the marine mammals, otters are one of the most social, floating in groups, from a few to hundreds, known as rafts. To our collective amazement, another sea otter grasps the hand of our sleepy otter, a display not found anywhere else in the animal kingdom. Is it some type of bonding? An act of intimacy? In a manner of speaking yes, since the hand-holding decreases their chances of drifting away.

While my visit with the sea otters off the Monterey coast was a glimpse into their fascinating lives, marine mammals have intrigued humans for thousands of years. Aristotle recognized that whales are mammals, not fish, because they nurse their young and breathe air like other mammals.

The sea otter (*Enhydra lutris*) is one of a number of marine mammals, including cetaceans and pinnipeds, which are found along the western North American coast and depend upon the diverse habitat and abundant food supply of the California Current System. At one time, the ancestors of marine mammals were terrestrial species, living solely on land. Approximately 70 million years ago, structural, physiological, and morphological changes in these species started to occur; they evolved to a life in the water, where they could avoid land predators and take advantage of the immense and richly productive marine environment.

The earliest known fossil remains of marine mammals are those of whales, dating back to the Eocene Epoch, some 60 million years ago. In comparison, the earliest fossil seals are considerably younger, from 20 million to 25 million years ago (Miocene Epoch). The sea otters, closely related to present-day land and freshwater animals, have taken to the sea relatively recently, having inhabited the northern coasts of the Pacific Ocean since the Pleistocene, about 1 million to 3 million years ago.

Of all the marine mammals, sea otters are the smallest, slowest swimming, and least streamlined. Though they are occasionally found on land, they retain essentially no ties to the terrestrial environment because they are able to breed and raise their young entirely at sea. Interestingly, though they are not as morphologically modified as other marine mammals, sea otters have evolved and adapted to a completely marine lifestyle.

Along the west coast of North America, the historical range of the sea otter extended from central Baja California northward to Alaska. The first known account we have of Pacific sea otters is from archaeological data collected on the Channel Islands in Southern California. For thousands of years, it appears that the Chumash coexisted with the sea otters. The Chumash hunted sea otters for their pelts, which were used to make blankets, robes, and capes, but this subsistence hunting appears to have had little impact on the vitality of the sea otter population. For Native Alaskans, all animals, whether harvested for food or clothing, deserved respect, and none more so than the sea otter. The sea otter was known as "the brother," and among some tribes, only elders or expert hunters could wear the fur of the sea otter.

The Chumash not only hunted sea otters but also collected mussels, abalones, sea urchins, and other shellfish and fished the kelp beds around the Channel Islands. Little did they realize that their coexistence with productive sea otter, shellfish, and fish populations was one of a balanced and harmonious relationship. Disruptions to the surrounding complex marine ecosystem were minimal; the overall structure of the biological communities remained intact.

But European and Asian expansion into the Pacific region would have dire consequences for the sea otter population. In 1741, Vitus

Sea otter (worldwildlifewonders/Shutterstock.com)

Bering sailed from Russia to explore and map the coastlines of the North Pacific. On board the expedition was the German naturalist Georg Wilhelm Steller, who studied and documented the flora and fauna, including the sea otter, of the region. In 1751, Steller would publish the first scientific description of the sea otter in his book *De Beastiis Marinis* (*The Beasts of the Sea*). Bering, on the other hand, took a pragmatic view of sea otters. Enthralled by their thick, luxurious fur, Bering obtained 800 pelts, thus starting the sea otter fur trade, which would spread like wildfire throughout the Pacific. The majority of the pelts that were taken back to Russia ultimately found their way to Chinese markets. To the Chinese, the beauty and high quality of these furs outranked even the famous Russian sable pelts. The wearing of sea otter fur as belts, capes, and trim on silk robes became a sign of wealth and prestige to the Mandarins. By the 1760s, a full-blown fur rush was on in Alaskan waters, a hunt that would decimate the sea otter population. When Bering and Steller sailed the Pacific, it was estimated that the sea otter population stood between 100,000 and 300,000 individuals—a high-water mark that would never be matched again.

Through the latter half of the eighteenth century, Russians intensively hunted sea otters throughout the North Pacific until the depleted numbers there warranted a push to the south. By 1812, the Russians had established base and hunting camps at Bodega Bay and the Farallon Islands in California. But the Russians would have competition in the sea otter market when the Americans, the English, and the French soon began hunting sea otters. The Russians and Americans formed an alliance. This Russian-American Company hunted with no regard for the future of marine mammals. The American merchants supplied the ships, and the Russians procured Native Alaskans to partake in the hunt. A joint venture between the company and the American captain Nathan O'Cain resulted in the highest known catch of otters in one year—over 9,000 pelts. With multiple nations all pursuing sea otters, the hunt decimated the sea otter population—in thirty-five years, over 100,000 pelts were taken. By the end of the nineteenth century, sea otters were at the brink of extinction, and trade in California sea otters came to a grinding halt. There were simply not enough sea otters remaining to make a profit. The surviving sea otters, which may have totaled no more than a few hundred, were scattered throughout their former range. In 1911, the International Fur Seal Treaty, which was signed by the United States, Great Britain, Russia, and Japan, halted all commercial hunting of sea otters. With the enactment of this conservation milestone and reintroduction programs, sea otter numbers rebounded, and in recent decades, California has seen a modest expansion of the sea otter population.

Today, we recognize the sea otter as a keystone species—a species that is important in determining structures and dynamic relations within communities. When a keystone species is removed from its environment, the system becomes out of balance and could potentially collapse. As predators, sea otters are vital in maintaining a balance within kelp forests—one of the most biologically diverse and productive marine ecosystems on the planet.

Tiered like a tropical rainforest with a canopy and several layers below, the kelp forest of the eastern Pacific is mainly composed of two brown macroalgae species: giant kelp (*Macrocystis pyrifera*) and bull kelp (*Nerocystis leutkeana*). The kelp forests of the Channel Islands are dominated by giant kelp, while farther to the north near Monterey Bay, both species coexist, but at times giant kelp will outcompete bull kelp for light. In the more northern reaches of the kelp's extent, the forests of the Gulf of the Farallones and the waters off of the Olympic Coast of Washington are comprised of predominately bull kelp.

Regardless of its composition, kelp forests are cool-water communities, thriving in water temperatures less than 60°F. They are most prolific in coastal upwelling regions, where cold, nutrient-rich bottom waters bathe the vegetation. Specifically, kelp need an abundant supply of dissolved inorganic nitrogen, which is available in cold, upwelled waters. In contrast, warmer waters tend to have low inorganic nitrogen because what was there gets quickly used up and is not replaced. In an experiment conducted near Santa Catalina Island, California, an adult giant kelp was transplanted from an inshore kelp forest to an offshore, low-nitrogen environment. For the first two weeks, the kelp maintained growth on internal nitrogen reserves. But by the third week, the growth rate decreased almost 75 percent as nitrogen reserves were depleted. In contrast to another type of kelp, *Laminaria longicruris*, which is adapted to long periods of low nitrogen availability, giant kelp has a small nitrogen storage capacity. While internal reserves of giant kelp appear adequate to make nitrogen starvation relatively rare in Southern California kelp forests, these forests may show signs of stress and deterioration during the summer months.

Kelp survival is strongly linked to the nature of the substrate. A rocky bottom—the larger and heavier the rocks the better—is needed for the kelp to anchor itself. Instead of treelike roots, kelp has holdfasts that grip the rocky substrate. Once anchored, kelp exhibits one of the most remarkable growth rates in the plant kingdom. The kelp *Macrocystis* grows faster than tropical bamboo—about 10 to 12 inches per day. Under ideal conditions, giant kelp can grow an astonishing two feet per day. In one growing season, these plants may attain a height of 150 feet. Held upright by gas-filled bladders, or pneumatocysts, at the base of leaflike blades, kelp fronds grow straight up to the sunlit surface, where they spread to form a dense canopy.

A host of marine organisms, including invertebrates, fish, and marine mammals, exist in the kelp forest environs. From the holdfasts to the surface canopies, the array of habitats may support thousands of invertebrates. In the Pacific kelp forests, many of these invertebrates are grazers, which include a number of species of abalones, limpets, and sea urchins.

Studies have shown that the preferred diet of sea otters includes not

Kelp forest (Ethan Daniels/Shutterstock.com)

only large abalones but also sea urchins and crabs. With a high metabolic rate, two to three times that of a comparatively sized terrestrial mammal, the sea otter makes a number of foraging dives to the seafloor during the day. The dives are usually short in duration, one to three minutes, but while submerged it is highly efficient in collecting food items. Using its front paws, it deftly lifts and turns over rocks—a feat no other marine mammal is capable of doing—searching for prey. To dislodge an abalone from its rocky perch, the otter uses a large stone, hammering away at the shellfish until it is dislodged. Thirty or forty blows and multiple dives may be necessary to dislodge the abalone that clings tenaciously to its bit of the sea bottom.

While less effort is needed to procure a sea urchin, care must be taken. To eat this spiny, hard-shelled animal, the sea otter flips over the urchin, bites through the underside where the spines are shortest, and licks the soft flesh from the shell.

Of all the invertebrate grazers within the kelp forest, the sea urchin is the most voracious, with kelp one of its main forage items. When sea otters are present, they keep sea urchins in check, restricting them to crevices where they feed on drifting kelp fragments. Urchins feeding in this manner are usually sedentary and simply wait for the material to come to them. But when sea otters are eliminated from the kelp environment, urchin populations are free of predation pressure. Expanding in numbers rapidly, urchins rove about the seabed, feeding directly on growing kelp, and over time denude the bottom of all potential new recruiting kelp. A three-dimensional vibrant kelp forest is essentially transformed into a two-dimensional urchin barren or bare bottom.

Archaeological evidence suggests that the kelp forests of the Aleutian Island chain experienced two distinct cycles of collapse on account of the intensive hunting of sea otters and the subsequent explosion of sea urchin populations. Intensive sea otter hunting in historic times by Russians led to extensive deforestation of kelp beds that once ringed this island chain.

In contrast, Southern California kelp forests did not experience wholesale collapse in the 1800s, even with the near eradication of sea otters in the area. The biological complexity of these forests is greater than their Aleutian counterparts, with a more diverse food web and a greater array of urchin predators, including the California sheepshead and spiny lobster.

With the historic extirpation of California sea otters and the collapse of the foraging economies of the Native Americans, the biological communities within the kelp forests underwent a significant change. Anomalously large populations of abalones, sea urchins, and lobsters took hold and became the dominant species. For a while, lobster and sheepshead predation and competition with abalones kept the urchin population in check. But a tipping point would occur with the advent of commercial and recreational fishing for abalones and lobsters. Extensive fishing pressure depleted these species to such an extent that the sea urchin population bloomed, resulting in deforestation of once healthy Southern California kelp beds.

In recent decades, tensions between government officials, fishermen, and environmentalists have risen due to the decline of the once-profitable abalone and sea urchin fisheries. As the number of otters increased—currently there are about 2,500 otters—along the Central California coast, their impact on shellfish beds was catastrophic for the commercial shellfishery. Biologists estimate that 100 sea otters can consume a half-million to 1 million pounds of abalone, lobster, crab, and urchin per year. Scale that up to the size of the present-day population, and sea otters are consuming 10 to 20 million pounds of shellfish on an annual basis. And sea otters live fifteen to twenty years.

Fishermen feel that they have been betrayed by resource managers who originally established an "otter free" zone (all waters south of Point Conception) in 1987 to protect California fisheries. The U.S. Fish and Wildlife Service (FWS) was charged with capturing all sea otters in the management zone and relocating them to San Nicolas Island in the Channel Islands. This site was selected because of its relative isolation, about seventy miles offshore from Los Angeles, and because it was within the historical range of the southern sea otter.

In principle, the purpose was to decrease competition for a limited resource and reduce sea otter mortality from entanglement in fishing gear. But to the dismay of the fishermen, FWS refused to remove any more otters, citing evidence that capturing and releasing otters proved to be more difficult than anticipated. In addition, translocation efforts were fatal to some sea otters, and some otters simply returned to their original location. FWS biologists and other independent researchers also believe that the California sea otter population, though increasing in range and size, is not a stable one but rather subject to perturbations, such as habitat loss and degradation, oil spills, and disease, which could cause a downward spiral in otter numbers.

Conservation and environmental groups entered the fray, claiming that the existing management zone and translocation program violated the Endangered Species Act under which the sea otter is listed as a threatened species. Casting even a bigger legal net, these groups argued that the sea otter was also protected by California state law and the federal Marine Mammal Protection Act (MMPA). In essence, the MMPA strictly prohibits the "take" of marine mammals in U.S. waters, where "take" under the MMPA is defined as "harass, hunt, capture, kill, or collect, or attempt to harass, hunt, capture, kill, or collect."

With biological and legal arguments on its side, the FWS, at least for the foreseeable future, decided to terminate the California translocation program after a twenty-four-year run and suspended any further capture of sea otters found in the management zone. As Ron Lohoefener, regional director of the service's Pacific southwest region, pointed out, "We've learned a lot during the course of the translocation program, and as a result have fundamentally changed our recovery strategy. Our experience strongly suggests that the best course of action is to allow sea otters to expand naturally into Southern California waters." But as of the summer of 2013, FWS's decision was being challenged by the California Sea Urchin Commission, an organization that represents urchin harvesters, and it has taken the traditional view that sea otters are competitors.

While this battle may ultimately play itself out in the courts, it might be valuable to take a historical perspective on resource management. We again turn to the Chumash and their predecessors, who collected shellfish from Channel Island waters for almost 12,000 years, fished the kelp beds for 10,000 years, and hunted sea otters for at least 9,000 years without upsetting the ecological balance. Some have viewed it as a very early case of the successful management of marine resources by the Native population. Unlike the foreign contingent of hunters who followed them, the Chumash appear to have hunted sea otters for millennia without causing the otter population to crash. Anthropologist Jon Erlandson, who has extensively studied and documented the human impacts on ancient environments, proposes that the Chumash may have intentionally culled the sea otter population to enhance the local shellfish harvests. If so, Erlandson believes that it might be worthwhile to consider applying a Chumash-style management model to current resource issues. While not advocating the killing of sea otters, Erlandson does believe that otters need to be excluded from those areas that historically have been productive shellfish and urchin fisheries.

A contrary view to Erlandson's is that the Chumash were not as skilled maritime hunters as were Alaskan Aleuts, who were responsible for the local depletion of otters. So the "management" by the Chumash was due to their limited technology, not by design. They could not have exterminated the otters because they lacked the expertise to do so.

In the near future, it appears that the conflict between fishermen and otters will continue given that, as with most marine populations, the availability of food is the main factor limiting the size of the California otter population. The population can grow only if the otters extend their range, wandering upon new food resources.

At present, the range of the southern sea otter is significantly compressed compared with historical times, extending from San Mateo County in the north to Santa Barbara County in the south. As food competition increases, young, subdominant male sea otters are forced to the limits of their range. These peripheral male groups, commonly known as the migrant front, are first to colonize new food-rich areas.

But at the heels of this vanguard are the older, dominant males, with which the young otters cannot compete, forcing them to keep pressing toward the edge. Eventually females, some with their pups, spread into the area, the population builds, food levels are depleted, and the cycle repeats itself.

Hunger is a strong motivator. Some more-adventurous sea otters may wander many miles ahead of the migrant front, but these wanderers rarely remain in one area for more than a few weeks. For resource managers charged with balancing the often conflicting interest of fisherman and conservationists, the range of the otters is the area between the fronts and does not include any new area temporarily staked out by those otters out ahead of the front.

The future of a vibrant sea otter population along the Pacific coast is inextricably tied to the vitality of the cold-water ecosystem in which the group resides and is an integral part of. With kelp being the foundational organism within this ecosystem, it is imperative to view its long-term stability, which has been and is impacted by anthropogenic and environmental factors. Recent work in paleoecology and paleoceanography has indicated that today's Pacific kelp forests are less extensive and lush than those in the relatively recent past. Since the end of the last ice age (20,000 years ago), kelp forests have flourished, tripling in size to their peak about 7,500 year ago. They then shrank by up to 70 percent to present-day levels. This dramatic response is attributed to climate forcing, particularly rising sea levels as the result of the warming trend beginning at the end of the glaciation. Understanding the past history of how a population responds to external pressures is crucial to providing guidance for assessing the impacts of a changing climate.

Potentially the greatest threat to the long-term stability of kelp forests comes from the commercial harvesting of kelp, which dates back to the turn of the twentieth century. About 100,000 tons of giant kelp are harvested per year, primarily for alginate, which is used as a binding agent and emulsifier in a broad range of health and pharmaceutical products and as a food source in abalone aquaculture operations. Currently, the area and volume of kelp harvest is strictly monitored and regulated by the California Fish and Game Commission, and due to its vigilance, it appears that the harvest of giant kelp is a sustainable venture.

Conditions are different with bull kelp. A growing concern exists that to meet market demand the bull kelp harvest will increase markedly along the Oregon and Washington coasts, where much of this kelp species resides. In addition to sea otters, many species, including shorebirds, juvenile salmon, rockfish, and shrimp, within the California Current–upwelling complex rely on bull kelp forests for feeding, spawning, and habitat. A recent study sounded the alarm that bull kelp is more sensitive to extraction than is giant kelp, a troubling finding that could lead to overexploitation of this species and a cascading effect throughout the kelp forest food web.

Pinnipeds

It was the great Swedish scientist Carolus Linnaeus (1707–78)—his life devoted to developing a taxonomy for classifying plants and animals—who placed seals, sea lions, and walruses into the order he called pinnipedia, from the Latin words *pinna* (feather or wing) and *pedes* (feet). His classification was a perfect representation of these marine mammals that have front and rear flippers, which they use mainly for propulsion. Taxonomically, pinnipeds comprise three families: phocids (seals), otariids (sea lions and fur seals), and odobenids (walruses).

Modern pinnipeds most likely evolved from "bearlike" ancestors during the early Miocene, apparently becoming aquatic in nature to take advantage of new food resources created by cooling ocean waters that resulted from a major climatic change about 36 million years ago. In the scientific community, there has been considerable debate among evolutionary biologists about the origin of pinnipeds. Today, the consensus of opinion, swayed by recent molecular and morphological data, is that phocid seals and otariid seals diverged from a single common ancestor about 25 million years ago, referred to as a monophyletic origin. Appendages, such as limbs and feet, have evolved over time into fins and flippers.

For the Inuit Eskimos of Alaska, there is no doubt about the origin of pinnipeds. The story that has been told for centuries revolves around Nuliajuk (also known as Sedna the Eskimo seal spirit)—a powerful being who was the mother of all animals. But Sedna was at one time a mortal. As the story goes, Sedna is thrown overboard during a storm by her panic-stricken father, who mistakenly believes this act will appease some seabirds that he has offended. As Sedna desperately clings to the side of the boat, her father cuts off her fingers, which are transformed into seals. Ever since, Sedna, or the spirit Nuliajuk, has had complete dominion over seals, taking them away from hunters when she is angry and returning them on a whim.

Other stories of seals and humans can be found throughout history. In Greek mythology, Proteus, an old sea-god, was the trusted herdsman of the seals of Poseidon. It was said that he took a daily nap with his ungainly herd on the island of Pharos. The seal-folk of Norse and Celtic folklore, variously called selkies or selchies, were believed to live as seals in the sea but shed their skin to become humans on land. While on land, they often interacted with other humans, marrying them and bearing their children. But they could return to the sea as seals again by donning the magical skin.

Not everyone thought of seals as mythological characters, possessing unique traits and powers. Linnaeus describes pinnipeds as a "dirty, curious quarrelsome tribe, easily tamed, and polygamous." With regard to the latter trait, Linnaeus may not be far off, particularly when applied to elephant seals. One alpha male can control a territory containing a cluster of females. As a result, a male northern elephant seal may impregnate more than forty females in a season.

On Land and at Sea

There are six species of pinnipeds that can be found throughout the California Current system: California sea lion (*Zalophus californianus*), Steller (or northern) sea lion (*Eumetopias jubatus*), Pacific harbor seal (*Phoca vitulina*), northern elephant seal (*Mirounga angustirostris*), Guadalupe fur seal (*Arctocephalus townsendi*), and northern fur seal (*Callorhinus ursinus*). (Walruses are limited to the Arctic waters of the Atlantic and Pacific.) Of these pinnipeds, two—the Pacific harbor and the northern elephant seal—are sometimes referred to as earless seals or "true" seals. These seals have ear holes but no external ear flaps. Thus, members of the otariids are often known as eared seals; phocids are sometimes called earless seals, lacking any visible outer ears.

Another morphological difference between the two groups is the nature of their rear flippers and their use for locomotion. The hind flippers, which are covered with hair, of the phocids cannot be turned forward under the body, severely limiting their ability to move about on land. They are almost comical in appearance as they slowly and awkwardly bounce and lurch forward. All otariids, in contrast, have almost

Male elephant seals (Richard Fitzer/Shutterstock.com)

hairless rear flippers that can be pivoted forward and used for movement on land.

But in the sea realm, even the most cumbersome pinniped on land is a swift and agile animal. Phocids rely on their rear flippers and lower body movement, side to side, to propel them through the water. Otariids use their long, front flippers like oars to swim effortlessly through the water. Their rear flippers do enter into locomotion but mainly act as stabilizers.

As a group, pinnipeds are not as streamlined in appearance as cetaceans. While some cetaceans, such as killer whales, can reach speeds of thirty miles per hour, most pinnipeds top out at about fifteen miles per hour, but higher, shorter bursts of speed have been documented in sea lions. Because sea otters feed on sessile or slow-moving invertebrates, as compared with swifter fish and cephalopods that are consumed by pinnipeds, they swim at the slowest rate of all marine mammals, approximately two miles per hour.

While pinnipeds are definitely at home in the water, they frequently wander onto to land—a vestige of their terrestrial origin millions of years ago. They can be found lounging, preening, socializing, and breeding on sandy beaches, islands, and rocky outcrops all along the Pacific coast. Pinnipeds are unique among mammals in that they couple a primary aquatic existence with the bearing of offspring on land. At times, five species make the Farallon Islands home, including the northern elephant seal, harbor seal, Steller sea lion, California sea lion, and northern fur seal. Even tiny San Miguel Island, part of the Channel Islands, may host all six species of pinnipeds, five of which will breed on the island. At any time of the year, one may find between 1,000 and 10,000 animals using the island. The elephant seal, in particular, appears to be expanding its historical range, moving northward over time and developing new breeding colonies on Castle Rock off Crescent City in Northern California and Shell Island off Cape Arago in Oregon.

No space is too small. A navigational buoy might be a good place to see a congregation of seals that have hauled out of the water for a bit of respite. Sail out past the Monterey Coast Guard jetty and be greeted by the loud "barking" of California sea lions that have found a temporary home there.

The Big Chill

Surviving immersion in cold water is no small feat for any mammal. The thermal conductivity of water is twenty-five times that of air; in other words, water will draw out heat from the body of a mammal twenty-five times faster than does air. For humans, the survival time is one to six hours for water temperatures between 50°F and 60°F but drops to between one and three hours with temperatures between 40°F and 50°F. And well before those times are reached, numbness and unconsciousness will occur. Conserving heat is therefore of paramount importance to all mammals who have entered the ocean. How do the pinnipeds maintain their core temperature of approximately 100°F?

For thermoregulation, size does indeed matter. A large size helps retain heat; large-bodied mammals lose heat less quickly than small mammals because the surface area where heat is lost is much less than the overall volume. In comparison with most of their terrestrial counterparts, marine mammals are relatively large—a distinct advantage for those that reside within the cold waters of the California Current Ecosystem. Elephant seals, the largest of the pinnipeds, reaching weights of over two tons, are able to spend much of their time diving due to their bulk. Pinnipeds also have a relatively compact body shape with no long limbs, other than flippers, from which to lose heat.

As terrestrial mammals invaded the aquatic environment over the course of geologic time, this transition was accompanied by the development of specialized insulation to aid in retaining body heat. Biologists believe that the overarching factor that allows marine mammals to maintain a high, stable core temperature is their insulation. For most of these ocean dwellers, the insulation is either fur or blubber—the subcutaneous layer of fat below the skin—or both. As mammals became more specialized for living in water, the reliance on external insulation, such as fur, decreased, while a thick blubber layer became more prominent. Although fur is a better insulator, particularly in air, than blubber and is also considerably lighter, its effectiveness as a thermal barrier is compromised in water because, during a dive, the insulating air layers between the hairs are compressed by hydrostatic pressure. In contrast, blubber is not compressed appreciably by water pressure; thus its insulating properties are not affected during a dive.

For totally aquatic mammals, such as whales, the blubber is the primary form of insulation. In contrast, sea otters, the most recent of the marine mammals to inhabit the sea, rely on a thick, waterproof fur for insulation. The densest fur in the animal kingdom, ranging from 750,000 to 1 million hairs per square inch, keeps the otter warm, even though its lacks insulating blubber. The fur consists of two layers: long, waterproof guard hairs and short underfur. The guard hairs keep the dense underfur layer dry, keeping heat-sapping cold water completely away from the skin. In addition, most sea otter dives are shallow, limiting compression of the trapped air layers.

Pinnipeds are the only marine mammals to retain both types of insulation. Although the phocids have retained a hair covering over time, they rely primarily on a thick blubber layer due to the poor insulating quality of hair that has become completely soaked. The otariids employ two different strategies for keeping warm while they forage. Fur seals have both a thick layer of fur and a moderate blubber layer; while sea lions depend solely on their blubber to retain heat. Europeans upon first coming in contact with the northern fur seals of California named them "sea bears" due to their thick fur, which rivals that of sea otters in density. The name seemed to stick since their genus name *ursinus* means "bearlike."

On average, phocid seals have the thickest blubber layers, which are three to five times thicker than sea lion blubber. The fur seal, as might be expected, has the thinnest blubber layer. Those pinnipeds that have the thickest blubber layer have the thinnest hair layer and vice versa.

The natural selection pressures driving the evolutionary transition from fur to blubber in most marine mammals may not be solely due to blubber's thermal properties. Studies have shown that the blubber of some pinnipeds, in particular, the phocids and sea lions, is vertically stratified, with the outer and inner layers serving different functions. The outer layer functions primarily as an insulating barrier, and the inner layer, composed of different fatty acids, serves as an energy reserve, providing nourishment during long periods of fasting.

Pinnipeds will voluntarily forgo feeding opportunities when they haul out to breed. The timing of the haul-out does not necessarily coincide with decreased foraging opportunities but appears to be triggered by biological rhythms. In other words, fasting periods are a regular event, tied to the life history of the species, and are associated with mating and breeding.

Some pinniped males, such as the northern elephant seal, will not eat or drink for up to three months during the breeding season. While on land, their efforts are centered on maintaining a territory or competing for dominant rank within the breeding rookery. Leaving the land to feed is not an option for a breeding seal due to the fear of losing its mate to a potential rival. Most female phocids voluntarily fast while delivering copious amounts of high caloric milk to their pups. In spite of the fasts, activity levels remain high, fueled by the energy stores in the blubber. At the beginning of the breeding season, as much as 50 percent of the total body mass of a seal can be made up of blubber. In contrast, harbor seals are unusual in that they eat during lactation.

If blubber is superior to fur with regard to its overall insulating properties as well as energy storage, then why haven't sea otters and some furred pinnipeds shed their fur in favor of blubber? It appears that the evolutionary switch from fur to blubber is dependent upon body size. For small-bodied marine mammals, fur may be their only option as a thick blubber layer would comprise a disproportionately large percentage of the mammal's body size, thus severely limiting their flexibility and maneuverability both on land and on water. Take, for example, two members of the otariid family: the northern fur seal and the California sea lion. The generally small-bodied fur seal has retained its fur as the primary insulator, whereas the large-bodied sea lion has made the switch to blubber without sacrificing mobility. Small marine mammals that rely on fur as an insulator have limited diving capabilities; larger animals, with a thick blubber layer, are best configured for a fully aquatic lifestyle.

In mammals, the process of metabolism generates heat by burning fuel or food. Biologists have determined through extensive measurement that some pinnipeds, particularly the otariids, have a significantly higher rate of metabolism than phocids of similar size. If so, this would supposedly aid in their ability to stay immersed in cold water for extended periods of time. By comparison, phocids, such as the elephant seal, are well insulated and have normal metabolism. In other words, blubber will do quite nicely in warding off the cold.

Pinniped pups that are born on land will eventually hear the call of the sea. At first glance, they appear to be ill equipped to enter the water to feed on their own, born with very little blubber and their small size contributing to heat loss. But what the pups do have are higher metabolic rates compared with the predicted rates of adults of similar size. These rates are in line with those of terrestrial mammal young. The fuel for the pup's metabolism is the rich, concentrated milk they obtain through nursing. The milk of many seals is very high in fat content, 40 percent or more, compared with only 2 percent in humans, 4 percent in cows, and 17 percent in reindeer. In addition, milk aids the pup in putting on blubber. As the pup matures, the decrease in metabolic rate is offset by savings due to increased size and thicker blubber.

Conversely for fasting seals, it is actually advantageous to reduce their metabolic costs, particularly when fasting occurs simultaneously with energetically costly activities that are part of their life history, such as lactation, breeding, and postnatal development. Sea lions, for example, during their almost two-week fast, may lose approximately 12 percent of their body mass, and their resting metabolic rate decreases 31 percent, which is typical of a "fasting response."

A pinniped that is able to regulate and maintain a stable core temperature through a wide range of ambient temperatures is conferred a significant survival advantage. As a "warm-blooded," or endothermic, organism, a seal can pursue fast prey because the flow of blood into its muscles warms these tissues, resulting in increased efficiency.

After depriving themselves of food for three months during the breeding season, northern elephant seals have only one thing on their mind: food and plenty of it to offset the loss of nearly half their fat mass. From

California sea lion (vagabond54/Shuttestock.com)

the rookeries along the Mexican and Southern California coasts, they travel north into the cold Pacific to reach their foraging grounds in the North Pacific and Gulf of Alaska, one of the longest migrations known to humankind. The need to feed necessitates staying at sea for extended periods of time (250 days for males and 300 days for females)—a potentially daunting task that could stress lesser organisms. Blessed with a very low metabolic rate, elephant seals are able to dive for long periods as they hunt for food. Their thick blubber layer gives them excellent insulation to function in the cold water.

Pinnipeds and other marine mammals, and even some fish, such as the bluefin tuna, have a complex of fine parallel arteries and veins, known as the *rete mirabile* (Latin for "wonderful net"). This versatile network functions as an efficient countercurrent heat exchange. In a pinniped, warm arterial blood flows toward its flippers, which because they are thin and contain no blubber could be a major source of heat loss. Heat is transferred from the arteries to the adjacent cold veins in the flippers. The veins then return the warmed blood back to the body, thus retaining heat that aids the pinniped in maintaining a relatively constant body temperature during its prolonged dives.

Into the Abyss

Without a doubt, thermoregulation in pinnipeds has increased their foraging opportunities, opening up a three-dimensional world. But diving involves the need for a whole new set of physiological adaptations by pinnipeds and other breath-holding organisms.

In 1870, Paul Bert, a French zoologist, physiologist, and politician, published the first comprehensive review of diving in vertebrates, in which he discussed the relationship between diving duration limits in aquatic birds and mammals and their blood volume. By 1939, another study detailed the physiological adjustments to diving by marine mammals. And by the 1960s, attempts to decipher the mysteries of diving in marine mammals had become widespread.

What have we learned about these masters of diving? How can elephant seals, for example, dive to depths of over a mile, stay submerged for almost two hours, and withstand the great pressures at these depths?

What might first come to mind is the amount of oxygen that the organism is able to store before the dive. An absence of oxygen means anaerobic respiration becomes the only source of energy, but anaerobic respiration produces only 5 percent of the energy compared with aerobic respiration. Such a great deficit can damage brain cells and have fatal consequences. When anaerobic processes dominate, lactic acid is also produced as a waste product, which can ultimately cause fatigue, resulting in decreased diving and foraging abilities. For marine mammals, their most efficient diving is done aerobically.

The main stores of oxygen in mammals are the lungs, blood, and muscles. Seals differ from humans in that they carry most of the oxygen they need in their blood (in hemoglobin) rather than in their lungs. Having a large store of oxygen in the lungs can be problematic for adept divers. Increased buoyancy results when the lungs are filled with oxygen, making it more energetically costly for the animal to dive. Before a dive, elephant seals will actually exhale to reduce the lung volume even more, thus decreasing the required energy. Lungs, as gas-filled spheres, are susceptible to the high hydrostatic pressures that are encountered on deep dives, potentially leading to their collapse.

Phocids have a much higher oxygen capacity than that of sea lions and even small whales. The oxygen store in the blood may be two to three times higher in these seals, such as the elephant seal, than in other pinnipeds. But even with high oxygen stores, the rigors of the dive would soon deplete the available oxygen, and the organism would reach its aerobic diving limit (ADL)—the maximum dive duration before there is buildup of postdive blood lactate concentrations. Diving within the ADL is preferred, since marine mammals would need extra time to recover from anaerobic dives—to rid their body of lactic acid—and therefore compromise their overall foraging efficiency.

During a dive, a number of complex vascular responses occur to conserve oxygen, one of most important being a reduction in heart rate, known as bradycardia. A seal resting on a beach has a heart rate of 55 to 120 beats per minute, but upon diving, the heart slows to 4 to 15 beats per minute. Despite this drop, blood pressure remains steady on account of decreased flow of blood to the extremities, allowing the blood to flow primarily to vital organs and the brain. Vasoconstriction may occur over 80 to 90 percent of the body. This redistribution of blood during the dive is another key to extended breath-holding ability, promoting the parsimonious use of the limited oxygen store.

Even though a number of marine mammals are capable of long dives, most of their dives are actually short and aerobic. Why? A seal that has just surfaced, say from a 50-minute anaerobic dive, may need a recovery period of 75 minutes to return to predive lactic acid concentration. The muscle fatigue resulting from the buildup of lactic acid renders the seal incapable of further activity for some time. Thus, out of a total of 125 minutes, only 40 percent of the time is spent diving. In contrast, a seal that dives six times but stays submerged for only 15 minutes per dive requires only 4 minutes of recovery per dive—80 percent of time is spent underwater.

Female elephant seals, for example, make good use of their time while at sea, diving repeatedly for food. They rarely spend more than a few minutes between dives. The amount of food a female is able to find on these foraging dives directly affects her breeding success and, if she gives birth, her pup's growth rate and survival. While most of the seals will feed well offshore, migrating to the North Pacific Transition Zone over an eight-month period, smaller numbers of seals feed in coastal regions, pursuing bottom-dwelling prey along the continental shelf. Their dives are relatively shallow and short, thereby maximizing their foraging opportunities.

The major difference between diving mammals and scuba divers, who are most likely to suffer decompression sickness, is that the latter breathe compressed air at depth, whereas marine mammals breathe air at atmospheric pressure. But while marine mammals perform single deep and long dives without experiencing any symptoms of decompression sickness, extensive foraging bouts may result in nitrogen gas accumulating in tissues, increasing the risk of decompression sickness. (Symptoms of decompression sickness have been observed in commercial free divers who routinely make long, deep, and frequent dives.) And yet, marine mammals typically do not get the bends.

The main defense mechanism against the bends is the mammal's lung structure, which collapses under high pressure. This forces the air away from the alveoli and into the upper airways where the gas cannot enter the bloodstream, preventing the blood from absorbing too much nitrogen at depth. In addition, the same diving reflex—decreased heart rate and increased peripheral resistance to blood flow—that allows pinnipeds to conserve oxygen also appears to explain why these organisms avoid decompression sickness upon ascent from the abyss. Since dissolved nitrogen bubbles in tissues are the main culprit, limiting the amount and distribution of nitrogen appears to be the primary method to combat decompression sickness.

But a 2011 study suggested that diving mammals may indeed experience the precursors of decompression sickness under certain conditions. A postmortem analysis of beached whales, dolphins, and seals found symptoms of bubble formation, showing that their defense mechanisms against decompression sickness do not always function properly. The authors suggest that environmental factors, such as cold water, could force the mammals to balance other physiological needs (increased circulation for warmth) with the need to dive safely.

Pinniped Sensory Systems

Since pinnipeds are the most amphibious of all mammals, their sensory systems need to function efficiently both above and below the water's surface to support a wide range of behaviors, no small task given that such systems operate in vastly different physical environments. While diving, staying submerged, and returning unscathed to the surface are necessary components for successful foraging, the development and the refinement of various sensory systems have enhanced the ability of the pinnipeds to detect prey that can be quite mobile and elusive. Pinnipeds are unique in that they must adapt to light levels ranging from bright sunlight when they are on land to the almost total darkness found in the sea. The structure of the eye must be able to withstand large changes in pressure, and the optical system must compensate for differences in the refractive index between air and water.

The ability of an animal to see depends in part on a method of forming an image. Most mammals, including humans, rely on a lens and cornea to focus an image on the retina. But recall the last time you were underwater with no mask or goggles and tried to focus on an object. It was most likely quite blurry. This problem results because the cornea, having the same refractive index as that of water, is essentially useless underwater, no longer able to focus an image as it did on land. Wearing goggles or a mask restores the air-water interface, allowing you to see more clearly. Pinnipeds must rely solely on a highly spherical lens to focus the light into a clear image while underwater. And yet, pinnipeds are also able to see well in air, as evidenced by their ability to catch an object, such as a fish, thrown by a trainer. The exact optical mechanisms that are in play here are still under debate and may be different even between various species of pinnipeds. The visual acuities of seals and sea lions are in the range of a few minutes of an arc or similar to that of terrestrial species, such as dogs and cats.

Particularly challenging to the pinnipeds is the underwater environment where dimly lit conditions are the norm. To gather in the ambient light, seals have large, round eyes in relation to body size and pupils that dilate to big orbs. Moreover, they have a large number of photoreceptor cells, known as rods, in the retina that respond well to low light, particularly valuable when they are diving. In a further adaptation to low-light conditions, each of the eyes of pinnipeds has a well-developed tapetum—a specialized layer behind the retina—which acts like a mirror reflecting light that passes through the retina back through it a second time, essentially doubling the light-gathering capability of the rods. Many terrestrial mammals, especially those that hunt at night, as well as other marine mammals that are nocturnal foragers possess a tapetum, which makes the eyes of the pinnipeds, like that of a cat, appear to glow when a light shines on them.

In the monochromatic blue world of the sea, is there any advantage in being able to distinguish colors? The most commonly claimed advantage of having color perception is that it enhances contrast, thus increasing the visibility of an object against its background. Also, specific color signals allow an observer to determine in more detail something about the nature of the object.

The ability to distinguish colors in bright light depends on the number and type of color photoreceptors, known as cones, in the eyes. Most humans have three types of cones—S, M, and L—which allow us to perceive the full spectrum of colors. (The letters refer to the specific wavelengths at which cones absorb light: S for short wavelengths, as in the blue part of the spectrum, M for medium, and L for long.)

Most terrestrial mammals exhibit a dichromatic color vision pattern, based on having S- and L-cones (commonly in the blue and green portion of the light spectrum). Although some early behavioral studies of pinnipeds demonstrated that they were able to distinguish color, recent physiological and molecular genetic analysis of the cone receptors in a large sample of species show that seals and whales generally lack the blue cones and possess only green cones. This result appears to be a direct challenge to the behavioral finding because cone monochromacy is usually strongly associated with color blindness. Even more interesting, the loss of blue cones seems a rather poor adaptation to the blue-dominated light realm of the open ocean. But while pinnipeds may lack cones, they possess rods that are most sensitive to the blue-green light in their environment. Even if color vision is not an issue in the blue underwater environment, the S-cones would be better suited than the L-cones to detect intensity and contrast hues.

If underwater light conditions are far from optimal or the seal is not picking up visual cues, it may have to rely on its auditory sensitivity. A seal must be able to distinguish and decipher important sounds from the considerable amount of background noise in the ocean. The sea is far from silent, buzzing with all types of racket: from the thunderous explosion of a whale breaching the surface to the barely audible clicks, snaps, and crackles of myriad tiny organisms.

The detection of prey by pinnipeds is made possible by well-developed underwater directional hearing that picks up prey's vibrations transmitted through the water. Although the detailed mechanism for hearing underwater remains unknown, in water, sound waves do not reach the inner ear through the auditory canal; instead, the sound waves reach this organ through the skull but also come from all directions at once. The acoustic information that the pinniped receives is essentially useless because the mammal cannot pinpoint a sound source. As a compensatory mechanism, the inner ear is partially detached from the skull, thus significantly diminishing the nondirectional sound waves. Phocid seals exhibit additional morphological and structural modification to their auditory system to enhance sound reception and direction-ality. These distinctive changes have recently led to the description of a unique mammalian ear—the phocid ear.

In the water, some pinnipeds are able to hear over a wide range of frequencies (70,000 hertz). (In comparison, humans cannot hear frequencies over 20,000 hertz and below 20 hertz.) But not all pinnipeds are created equal with regard to their hearing sensitivity. Underwater, the elephant seal is most adept at detecting low-frequency sound emissions, followed by the harbor seal and the sea lion. Understanding what pinnipeds hear is not only important for relating it to their ecology but also for identifying species at risk from anthropogenic noise in the marine environment. Long periods of evolution have allowed marine mammals to adapt and tolerate the wide spectrum of natural acoustic signals that occur in the ocean, but "acoustic pollution" sources, such as sea traffic, sonars, underwater explosions, and geophysical measurements, may not only directly damage their hearing but also interfere with a number of their activities, including the ability to locate prey.

Dolphins, in particular, have evolved a sophisticated sonar system to explore their environment, particularly where the visual sense is of limited use. Known as echolocation, it depends upon specialized sound production, sound reception, and signal-processing mechanisms. Because echolocation provides a distinct advantage when foraging in dark waters and increases foraging opportunities, it has long been hypothesized that other marine mammals, such as the pinnipeds, possess an analogous biosonar. A high-frequency biosonar would be particularly advantageous because of the increased resolving power of a system using sound emissions with wavelengths smaller than the objects being targeted. The accumulated evidence, which was painstakingly gathered over decades, does not support the hypothesis of echolocation in pinnipeds. Recall that pinnipeds are amphibious, at home on land and in the water. A highly developed echolocation system most likely evolved only once in the ocean, in a group of marine mammal predators—the odontocetes (toothed whales)—which became totally tied to an aquatic existence early in their evolutionary development. Because odontocetes function totally in the marine environment, for example, never having to give birth on land, their acoustic systems became fully adapted for their underwater lifestyle, a refinement not possible in pinnipeds. Once hauling out on land, most pinnipeds, if not all, use airborne vocalization and hearing, primarily, as a means of communication in this social setting. While hearing in air and water requires different adaptations to accommodate the manner by which sound propagates through the surrounding medium, enters the animal, and ultimately reaches the primary receptor of the inner ear, no major tradeoffs in aerial hearing appear to exist that allow for adaptations for underwater sound detection. In other words, pinnipeds have not sacrificed in-air hearing capabilities for enhanced underwater sound reception. Auditory measurements have shown that harbor seals, for example, hear almost equally well in air and underwater.

Natural selection pressures that have favored the development and retention of in-air hearing may have, in turn, limited the formation of an active sonar-based system in pinnipeds for underwater orientation. The result is that pinnipeds have evolved to use three overlapping sensory channels for underwater foraging—visual, acoustic, and tactile (long, sensitive whiskers)—in lieu of echolocation, a scheme that has, over the eons, served the pinnipeds quite well.

Threats to Pinnipeds: Predators, Regime Shifts, and Humans

Because pinnipeds are amphibious, they have historically been subject to predation both on land and in the water. In the aquatic environment, their main adversary appears to be the great white shark, a species vividly on display in the movie *Jaws*. There are more than 350 species of sharks worldwide, of which only a few prey on pinnipeds, with *Carcharodon carcharias* (a Latin name derived from the Greek words for "ragged" and "tooth") probably the most skilled and stealthy hunter of pinnipeds. The great white is admirably adapted to killing and feeding. Its immense jaws are powerful, exerting over a ton of pressure per square inch and housing up to four full rows of granite-hard teeth. The lower teeth spike and hold the prey, while the triangular upper teeth with their sharp serrations can efficiently saw and cut through both flesh and bone.

One of the great white's main hunting grounds in the California Current Ecosystem is the Farallon Islands—a nesting site for thousands of seabirds and a safe haven for the hauling-out and breeding activities by pinnipeds. While prey is plentiful in the rich, fertile waters surrounding the islands, the dietary preference for adult great white sharks is seals and sea lions. Why might these sharks have a keener taste for pinnipeds than other prey items? Researchers have shown that sharks are

Great white shark (nitrogenics.com/Shutterstock.com)

opportunistic predators, feeding when and where food is available. But a number of studies have documented great white shark attacks that have not automatically led to the prey being consumed. In one specific attack, a shark seized a brown pelican, rendered it useless to escape, but refused to eat it even though the pelican was bleeding profusely. Also, in a number of instances, dead sea otters were found along the California coast with fragments of a great white's teeth embedded in them, but otherwise the carcass was wholly intact. The analysis of the stomach contents of a great white shark have yet to reveal the remains of a sea otter. The connection between these observations lies in the nature of the prey. Birds, sea otters, and, yes, even humans are composed mainly of muscle, whereas some of the pinnipeds have a thick layer of fat. And within the pinniped world, one of the favorites of the great white is the blubbery northern elephant seal, its thick insulating layer now a negative feature. Why does a great white have a preference for fatty prey? The answer lies in the physiology of the species. Great whites have one of the highest growth rates among sharks, twice that of a porbeagle shark, for example, which also inhabits cold temperate waters. To fuel its high metabolic needs and to hunt successfully in cold water, the great white needs the high-energy rich fat. Anything else is nutritionally inadequate.

The sharks travel to the Farallones in the fall, timing their visit with the arrival of juvenile elephant seals and sea lions. Shark attacks increase dramatically from September through November and are clustered at locations where seals enter or depart the water. In one season, more than eighty attacks were observed and recorded. Long-term observations of these attacks showed that they took place at relatively the same time on consecutive days. The timing of the attacks appears to be correlated with the occurrence of high tides that force the elephant seals from their safe beach haven into the shark-infested water.

Morphological differences seem to affect both the pinniped's appeal to the shark and its ability to survive an attack. Seals utilize their rear flippers for swimming; their small, underdeveloped front flippers are essentially useless for propulsion. In contrast, sea lions, which have strong, highly developed front flippers and less developed rear flippers, are attacked less frequently than seals and more likely to escape after the initial onslaught. As sea lions pass through the nearshore danger zone, they also employ a unique strategy to minimize the risk of attack: they swim in large groups and "porpoise" through the water. While the former probably confers the perceived advantage of safety in numbers, the latter is a bit more puzzling. One possible explanation revolves around the finding that great white sharks appear to select their prey by shape. Most shark attacks on pinnipeds occur from below, where from the shark's perspective the seal has a distinctive and familiar silhouette outlined against the light background. Does the shark possess a particular "search image" for seals, and does the swimming style of the sea lion distort this image? As is often the case in trying to interpret and understand behavior in wild animals, the answers to these questions must await more rigorous experimental study—a daunting task with such a volatile species as the great white.

For argument's sake, let's say that the porpoise-like swimming style exhibited by the sea lion is a successful evasive tactic. Is it instinctual, or does it reflect some higher cognitive process? Can we say that sea lions have found a solution to their problem? Pinnipeds tend to be intelligent and social animals, although such attributes vary according to species. Some groundbreaking experiments with captive sea lions have shown that, like dolphins and primates, they are capable of mastering complex tasks. And when visiting an aquarium or marine park, who has not been impressed by the tricks learned by sea lions? It has thus been argued that the evolution of well-developed cognitive and learning abilities among pinnipeds would presumably facilitate their existence in a complex and often bewildering environment. If a seal learns how to find and capture elusive prey, engages in complex social interactions during the breeding season, masters the intricacies of migration, and learns how to raise a pup, then it appears reasonable that the sea lion may have developed a new trick to outwit the great white.

But it may be a bad time to be a seal. That's because the numbers of great whites appear to be on the upswing. A recent survey estimated that 2,400 great whites now patrol the waters off the California coast, up from the previous estimate of 500 sharks. Some researchers argue that the increase is the result of decades of federal protection for marine mammals as well as the shark's designation as a prohibited species, meaning it cannot be targeted by fishermen. Others are a bit more hesitant to claim that the shark population has fully recovered. White sharks take years to reach sexual maturity; females must grow to about fifteen feet. Pregnancies are also long—twelve to eighteen months—and females give birth every two to three years to litters of two to ten offspring at a time. All of which means that if populations rebound, it happens at a relatively snaillike pace.

Though researchers employed a number of tactics to get a handle on the number of white sharks, in the end it was impossible to put an absolute number on how many great whites exist off the California coast. Direct counting of sharks would possibly alleviate any concerns about the animal's success, but this in no easy task. These elusive creatures travel great distances and rarely surface to be observed.

While the great white shark is the main aquatic predator of pinnipeds, an infrequent, but potent, visitor to pinniped rookeries, including the Farallon Islands, is the killer whale (*Orcinus orca*). Even an alpha great white is no match for this large and cunning mammal. Great white sharks can attain lengths of twenty feet and weight upward of 4,000 pounds. By comparison, killer whales can be thirty feet in length and weigh 13,000 pounds. On October 9, 1997, a great white found out just how formidable a killer whale can be when it was attacked and killed by the whale off the Farallones—the first recorded account of such an encounter. Even through prey was plentiful, the remaining sharks abandoned the Farallones, a visceral reaction to the whales?

Regardless of the reason, the Farallon seals and sea lions were not out of danger. Killer whales consume a variety of marine vertebrates, including pinnipeds. They are eclectic in their choice of prey: California sea lions, Steller sea lions, harbor seals, and northern elephant seals. As opposed to great whites that are essentially solitary hunters, killer whales often forage cooperatively—hunting like a pack of wolves—employing sophisticated and coordinated maneuvers to confuse and capture their prey. At sea, killer whale pods have been observed herding sea lions into a tightly packed group, surrounding them, and taking turns plowing through them to feed. The whales feed mainly on the pups, killing and consuming as many as twenty per hour. Like great white sharks, they are capable of consuming a large amount of food. In one report dating back to 1861, it was supposedly documented that the remains of fourteen seals and thirteen porpoises were found in the stomach of a twenty-one-foot killer whale. If true, the report speaks volumes of the prodigious appetite of whales for marine mammals, even though the prey items were probably consumed over a period of time.

Is there any evidence that this natural predation has an impact on the pinniped population? The ecological roles and effects of large carnivorous predators has not been adequately documented and quantified. This lack of information is not totally surprising since direct experimental verification is extremely difficult given, in part, the mobility, relatively low density, and elusive nature of these animals. But simulation and modeling methods, while not truly replicating the "real" world, have provided insight on the possible and potential impacts of large predators on prey resources.

One particular study by Terrie Williams of the University of California, Santa Cruz, and her colleagues focused on the killer whale because its large body size, carnivorous nature, and energy cost as a warm-blooded mammal require high caloric intake. The results were eye opening: fewer than 40 killer whales could have been responsible for the observed decline in Steller sea lions in the Aleutian archipelago, and only 5 individuals could account for the decline in sea otters and the continued suppression of sea lion numbers. Are these numbers truly representative of killer whale behavior? For an adult killer whale to meet its high energy needs, it was estimated that the whale, depending on its sex, must eat between 3 and 7 sea otters per day. Extrapolating that number in time, it follows that an individual killer whale focusing only on sea otters would consume 1,100 to 2,600 otters per year. In contrast, fewer Steller sea lions—2 or 3 pups per day or about 840 pups per year—

would be needed to meet the whale's energy requirements given their larger size and higher caloric content. (Steller sea lion pups, which have a blubber layer across their bodies, provide over 40 percent more calories per kilogram of body weight than do sea otters.)

If, as the study showed, increased predation is indeed responsible for the observed population declines in sea otters and Steller sea lions, then it is likely, as proposed by the researchers, that either a change in the killer whale population or behavior during the past three decades triggered the drop in numbers. Evidence seems to point to changes in foraging behavior as the most likely culprit because there is little evidence showing that killer whale numbers have increased substantially over this period. In particular, changes in traditional prey resources could have initiated a dietary shift, which may owe its origin to the demise of the great whales through intensive commercial hunting after World War II. Large whales have historically been an important component in the diet of killer whales. In the absence of this once-abundant resource, killer whales would have gravitated to seeking out other prey items to satisfy their high energy needs. The problem is exacerbated because the reproductive rates of the sea otters and sea lions do not counter the level of predation by the killer whales.

If the dietary switch by killer whales has indeed impacted specific pinniped populations (it should be pointed out that the study is somewhat controversial) and is continuing unabated even today, could the killer whales literally eat themselves out of house and home? Recent reports of transient killer whales feeding on harbor seals in Puget Sound, Washington, show that the predators may simply move into previously unexploited areas. The need to feed has pushed at least three killer whales (as documented by visual evidence) from Glacier Bay, Alaska, to central California, where they were observed attacking gray whales. If those same whales, as speculated by the study's researchers, had opted to feed on sea otters, the entire threatened population of almost 2,000 California otters could have been wiped out in less than four months.

A different perspective on the decline of the Steller sea lion, which has implications for other pinnipeds throughout the California Current Ecosystem, centers on their poor nutritional state, which appears to result from the absence of high-fat-content fish that have traditionally made up the bulk of their diet. The drastic decline of this prey has forced the sea lions to seek other options, such as pollack, which lack the necessary fat content to meet their energy needs. The question that is still unresolved is, what precipitated the shift? Climatic events, such as El Niño and Pacific Decadal Oscillation, can most assuredly alter the abundance of the food supply, but so can overfishing. Perhaps it is a combination of all these factors, or maybe something else is at work that has of yet not been detected.

What is not in question is that pinnipeds, like their sea otter cousins, have been hunted by humans throughout history, possibly beginning with Arctic Inuits over 4,000 years ago. As with sea otters, seals not only provided warmth and sustenance during the long, harsh winters but also were the linchpin for the Inuits' culture and customs. Elaborate rules governed the butchering and distribution of the seal meat. After killing a seal, the hunting party celebrated their good fortune by feasting upon the still-hot liver. Upon butchering the rest of the seal, the distribution of the meat was tightly controlled by kinship and allegiance. Partnerships that were formed often lasted the lifetime of the participants and were even passed down from generation to generation.

While seals were often prominently featured in Scottish culture, the Scots' attitudes toward seals were not necessarily reverential in the Middle Ages. The monks of the Isle of Iona Monastery harvested seals for their meat. Because the monks viewed the seals as "fish," they felt no qualms about eating them on holy days when consuming "meat" was strictly forbidden.

The human exploitation of pinnipeds during the nineteenth century "fur rush" in California was just as intense, if not more so, than the killing of sea otters. In addition to their fur, numerous species of pinnipeds were sought after for their blubber. The rich oil derived from the blubber was put to many uses: burned to provide light, used to lubricate engines, and employed to make soap and paint. More than 100 gallons of oil could be procured from an average-sized male elephant seal, and probably twice that from a very large bull.

So great was the onslaught on the pinniped population that in the Farallon Islands alone it took less than forty years to wipe out the seal species. From the initial encounter by an English sea captain in 1803 who wandered upon the huge colonies of sea lions and elephant seals on these rocky outcrops to 1840 when the colonies were gone, the hunting was relentless. In just two years, 1810–11, New England sailing boats slaughtered thousands upon thousands of northern fur seals.

Without a doubt, the pinnipeds' propensity to breed, molt, and rest on land made them easy targets to the men who invaded the numerous islands of the eastern Pacific. But why didn't the seals flee their pursuers? Was there any latent drive for self-preservation? Maybe they did try to elude their predators but their slow, lumbering gait proved to be an impediment. Or could it be that the pinnipeds had become conditioned to their surroundings—the crashing of the surf, the cacophony of nesting birds, or the presence of people—a type of habituation?

While local Native Americans assiduously avoided setting foot on the Farallon Islands, they most likely hunted and fished the fertile waters surrounding the islands and would have been in close proximity to pinniped rookeries. The presence and sound of these humans may or may not have frightened the animals into the sea, depending upon the species. Habituation to what is commonly referred to as the startle or flight response varies within different populations with respect to sex, age, and location. Many populations of sea lions in California, for example, shun any contact with humans, afraid of their mere presence. In contrast, the sea lions of the Galapagos Islands are rarely frightened by people, tolerating more human intervention, even curious about these other beings. Of all the species, harbor seals are the most skittish, constantly scanning their surroundings and even interrupting their sleep to assess potential threats. Other species, including the elephant seal, tend to be fairly tolerant of human presence and are the most easily approached pinnipeds. Unless it is breeding season when the females and bulls can be belligerent, humans get can close enough to almost reach out and touch these behemoths of the pinniped world. (A case in point in today's world is the yearly hunting of Arctic harp seals that, because they do not fear humans, are routinely bludgeoned to death by the commercial sealing industry.)

Even in the twentieth century, the pinniped colonies of the California Current Ecosystem were still under attack. In particular, poachers routinely raided sea lion breeding grounds and killed adults and pups for their fur, nearly wiping out the population of San Miguel Island in 1908. In the 1930s, the killing reached such a level—tens of thousands of sea lion bulls—that California officers feared it would meet the same fate of Steller's sea cow—extinction.

No issue has probably generated more controversy among conservationists, commercial fishermen, and biologists than the impact of pinniped predation on fisheries. Since some pinniped species eat many of the same species, including squid, salmon, herring, sardines, and anchovies, which fisherman target, the controversy has raged on for years.

The extant research has generated more heat than light on this topic. Some within the scientific community have argued that there is no evidence that pinnipeds, or any other marine mammal except for sea otters, are able to effectively diminish the catch by commercial enterprises. But tell that to a fisherman, struggling to make a living from the sea, who routinely comes upon pinnipeds poaching fish from his nets. In the early 1970s, angry and frustrated fishermen who caught sea lions stealing their catch often killed their perceived competitors on the spot.

The enactment of the Marine Mammal Protection Act in 1972 provided a blanket of protection to the beleaguered pinniped population, and their numbers rebounded. At present, about 300,000 sea lions now reside along the Pacific coast, from Alaska to Baja, California. The downside to this burgeoning population is that the sea lions are so numerous that they are close to their carrying capacity. These voracious predators have become even bolder in their encounters with commercial fishermen, and the fishermen have responded by going above the law. During 2009 and 2010, almost three dozen sea lions were treated for bullet wounds.

The decades-long conflict between these hungry predators and fishermen may have come to a head at the Columbia River in Oregon. These once-endangered species are now wreaking havoc on another species—the Chinook salmon. The impact has been significant. Biologists estimate that about 100 sea lions can consume thousands of Chinook salmon every year. In 2010, a good year for the sea lions, they devoured 6,000 salmon—more than 2 percent of all the fish. And during its seventeen-year lifespan, a sea lion may return every year to feast on the seasonal salmon run.

Some of the more audacious and ambitious sea lions have even traveled more than 140 miles upriver to the Bonneville Dam—a bottleneck for the salmon as they attempt to reach their spawning sites. The trapped and struggling salmon are easy pickings for the cunning sea lions, and their success is evident by their increased heft. A typical full-grown male may top out at close to 800 pounds, but at the dam, the animals are pushing 1,500 pounds.

The clash of one protected species against one endangered species makes for a tricky issue for officials to resolve. But an amendment in 1994 to the Marine Mammal Protection Act does allow for the possibility of limited removal of pinnipeds preying on salmon should the level of predation be determined to have a significant negative impact on the decline or the recovery of Endangered Species Act–listed salmonids. Specifically, according to National Marine Fisheries Service policy, "a sea lion is deemed predatory and marked for death if the animal is individually seen on any five days at the Bonneville Dam and is observed eating at least one fish on one of those five days." Up to ninety-two sea lions could be killed in any given year.

In October 2010, Washington and Oregon officials, after repeated failed attempts to relocate the sea lions, euthanized twenty-five sea lions by means of lethal injection. And the killing continues. As recently as April of 2014, six sea lions were put to death.

Some have argued, such as the Humane Society of America, that the killing is needless and that the sea lions are scapegoats for a salmon decline that has deeper roots than simply a sea lion doing what comes naturally—eating fish. In addition, those sea lions that are killed will simply be replaced by new recruits—a never-ending culling of a so-called protected species.

Chapter Six

CETACEANS
Whales, Porpoises, and Dolphins

These marine mammals have always caught the fancy of humans—from antiquity to the current whale-watching excursions—probably due to their huge size, perceived playfulness, value as a resource, and unique adaptation to a totally marine existence.

Whales and dolphins played a prominent role in the culture of the ancient civilizations that bordered the northern Mediterranean Sea. The Phoenicians were believed to have hunted whales in the Mediterranean as early as 1000 B.C. But some cetaceans, such as dolphins, were elevated to a mythical status in Greek society. When the god Apollo founded the oracle at Delphi, he took the form of a dolphin, on which he carried the holy priests to this site. Poseidon, the ultimate ruler of the seas, needed a wife who would be comfortable residing in the depths of the sea. He fell in love with the sea nymph Amphitrite, who at first spurned his advances. Undaunted, Poseidon sent his sea creatures, including Delphinus, to plead his case. So convincing was Delphinus that Amphitrite consented to the marriage, and as a reward, Poseidon set Delphinus's image in heavens as a constellation—Delphinus the Dolphin.

Even in the Bible another cetacean assumes a prominent role. In the oft-repeated story of Jonah and the whale, Jonah finds himself thrown into the sea during a great storm and is swallowed by a great fish, a sea monster, or as most generally described in the more recent versions of the Bible, a whale. For three days and nights, Jonah resides in the stomach of the whale, a penance for having disobeyed God, until he is mercifully ejected by the whale.

The Haida, an indigenous population of the Pacific Northwest, viewed killer whales as the most powerful creatures in the ocean, and their stories tell of killer whales living in towns and houses under the sea, where they took on human form. So highly were killer whales regarded throughout the Haida's art, history, and religion that it was believed that anyone who drowned would find a new life with the whales.

While legends, myths, and tales were the first mediums to thrust cetaceans into the limelight, the first detailed work given to the understanding of these marine mammals was Aristotle's *Historia Animalium*. In this treatise, Aristotle recognizes several distinct species of cetaceans, including dolphins, killer whales, sperm whales, and right whales. Moreover, he was careful to make a distinction between whales possessing teeth and those with "hairs that resemble long bristles" instead of teeth.

The Roman historian Pliny the Elder (A.D. 24–79) followed in Aristotle's footsteps with his exhaustive thirty-seven volumes of *Natural History*, of which volumes 7 through 11 painstakingly covered many aspects of zoology—everything from bugs to fish to whales. In particular, Pliny noted that some whales migrated to different places depending on the time of the year and during their journey were attacked by killer whales. Unlike Aristotle, Pliny believed cetaceans were fish. And yet his work, oddly enough, would effectively remain the factual text on cetaceans until the seventeenth century.

But some modern historians take a more circumspect view of Pliny's observations and documentation, accusing him of being negligent in fact checking and verifying what he wrote. Among the bizarre wonders he described were a monumental battle between an elephant and a dragon and a boy who rode to and from school on a dolphin's back. Science historian Brian Cummings simply described Pliny as "endearingly batty."

Over time, the study of cetaceans would spread from the Mediterranean region northward to the Nordic seas. By the ninth century, Norwegians had identified more than twenty-three different species of cetaceans that inhabited their waters. The Norwegian text *Speculum regale* (Old Norse for the "king's mirror"), written in the thirteenth century, makes note of a variety of whales, including sperm whales and narwhals, but also gives special attention to a species referred to as "orc" that had an aggressive nature and was disposed to attacking other marine mammals.

In 1970, a large Pacific storm exposed the remains of an ancient village of the Makah tribe, who inhabited Washington's Olympic peninsula. Archaeologists who sifted through the remains were astonished to find lances and harpoons that dated back to the 1400s. The Makah hunted whales that migrated along the coast—a journey that gray whales continue to make even today. The Makah would come to know a great deal about the habits and the natural history of the animals on which they depended for food.

With the advent of large-scale commercial whaling at the beginning of the seventeenth century to its heyday in the eighteenth century, the need to know the habits of these creatures was paramount to the success of the whaling industry. Whaling captains and merchants kept meticulous records of the movements of whales throughout the vast oceans, seeking answers to the questions of where they go and when would they return. Today, human interest in trying to decipher the secrets of cetaceans continues unabated, only not for hunting but for conservation.

One such ongoing effort is the Gray Whale Count, a research and educational endeavor in which land-based observers monitor the passage of gray whales as they migrate northward through the Santa Barbara Channel. These observers, or counters as they are officially referred to, may not always shout out "thar she blows" upon sighting a whale but have been vigilant in logging the number of gray whale adults and calves over more than a three-month period, a data set that will complement similar sampling studies along the coast.

Cetacean Evolution, Classification, and Diversity

The whales of today that swim the world's oceans, plumb the depths, and feed on the sea's bounty have an origin that is much longer than humans have existed on Earth. The oldest known whale fossil was unearthed from 50-million-year-old rocks in Pakistan. The fossilized skull, small compared with those of modern whales, belonged to the oldest group of whales, the primordial Archaeoceti. The little mammal, scientifically known as *Packicetus inachus*, was probably amphibious given that its closest relatives have been shown to possess front and hind limbs. *Packicetus* and its kind were almost crocodilian in appearance, having long, slim heads and pincerlike jaws often lined with serrated teeth. In addition, archaeocetes also had nostrils near the tip of the nose, like land mammals, rather than a blowhole on the top of head, like modern whales.

By the late Eocene Epoch (40 million years ago), the archaeocete whales had essentially abandoned their river and coastal haunts and had spread to many parts of the world's oceans, taking advantage of the new opportunities and niches provided by a dynamic and changing planet. In a blink of geologic time, new species found their way onto the world's stage. *Zygorhiza*, for example, had a whalelike body but retained the nose nostrils.

While Pliny the Elder may have viewed the idea of whales being mammals as a bit wild, he might have been surprised, as were some modern-day scientists, to learn that these giant aquatic beasts are closely related to the hippopotamus. One long-held belief within the scientific community was that hippos were related to pigs, or horses as the ancient Greeks thought, but modern phylogeny shows a close relationship with whales. But, even to the most casual observer, whales do not look anything like hippos, most assuredly unlikely cousins. And to complicate matters, there is a gap of tens of millions of years between the oldest known fossil of a hippo and that of a whale. This huge gap in the fossil record left the experts stumped. But a new study by Fabrice Lihoreau and his colleagues in France filled in that gap, finding the remains of a 28-million-year-old animal. This four-footed, semiaquatic mammal (*Epirigenys lokonensis*), which thrived for millions of years, is now recognized as the common ancestor to both whales and hippos.

The rise of modern whales occurred during the late Oligocene (30 million years ago) when two present-day lineages of cetaceans evolved from archaeocete ancestors. (The whales, dolphins, and porpoises all belong to the order Cetacea, which is derived from the Latin *cetus*, meaning "a large sea animal," and the Greek word *ketos*, meaning "sea monster.") It was a period of great change for the early whales, one marked by pronounced morphological changes. Ancient limbs had been reduced to mere vestiges, and skulls, necks, and teeth took on new appearances and form.

First to arrive on the scene were the toothed whales or Odontoceti. From this lineage would in time come the playful and intelligent porpoise, the fearsome killer whale, and the great sperm whale of Herman Melville's novel *Moby Dick*. Armed with an arsenal of teeth, these whales, depending, in part, upon their size, feast upon a variety of organisms: from fish to squid to other marine mammals.

The newest branch evolved into the baleen whales of today, which belong to the suborder Mysticeti. Baleen whales, such as the fin, blue, and California gray whales, have no teeth; instead, they have plates of a fringed horny material, called whalebone or baleen, hanging down from their upper jaws. Lining the plates, which are used by the whales to feed, are multiple strands of hairlike fibers. Like the teeth on a comb, these thick, fibrous strands allow water to pass through but trap any objects too big to fit between them, such as planktonic organisms and small crustaceans. Essentially, baleen whales are filter feeders, straining the water for their food.

Filter feeding, which is not found in any terrestrial mammals, arose in response to the unique spatial and temporal patterns of plant productivity and prey availability in marine ecosystems. In the ocean, due to differences in physical processes, the phytoplankton biomass is often patchy and ephemeral. Consequently, marine grazers, such as zooplankton and small fish, will often occur in extremely high densities in close proximity to these patches of phytoplankton. Filter feeding allows baleen whales to exploit this densely packed prey in a single feeding event, during which they consume copious amount of prey in the relatively short time. But for this foraging strategy to be effective, whales must travel, sometimes great distances, to where the prey is concentrated. A census of cetaceans conducted in 1979–80 indicated that these mammals were more abundant in the productive coastal waters than in the offshore waters of the California Current System. The distribution was linked to the high plankton biomass occurring in this region as the result of upwelling due to ocean events or geologic features, such as underwater banks and seamounts.

Long-term observations, for example, have shown that blue whales are aggregated in the cold, nutrient-enriched waters that have upwelled along California's Channel Islands during the summer and fall. Massive blooms of phytoplankton support a dense population of shrimp-like organisms that are exploited by the whales. Capturing and processing large quantities of prey in a single mouthful allows the whales to acquire energy at high rates when small prey items are concentrated. Interestingly, the late Oligocene whale *Aetiocetus* possessed skull and jaw features that are found in current baleen whales and is considered to be the earliest mysticete—even though it bore a full set of teeth.

By about the middle of the Miocene (15 million years ago), the whales of both lineages had become quite common, and the evolution of the modern whale was fully under way. And during the last Ice Age, or Pleistocene Epoch (1 million years ago), the world's oceans were populated by modern odontocetes, while the mysticete lineage continued to develop during this period to the present. What started out eons ago with an animal closely resembling a hippopotamus evolved into a mammal best adapted to aquatic life. The forelimbs have been modified into flippers, the body is streamlined, the tail has horizontal flukes, and tiny hindlimbs are vestigial, deeply recessed within the body. The great whales can never return to terra firma; their massive bulk could not be supported on land, and the lack of limbs would limit any mobility.

The odontocetes are the most numerous worldwide, comprising sixty-seven different species, as compared with eleven baleen species. Within the California Current Ecosystem, over two dozen of these species can be found.

What the baleen whales lack in numbers, they make up for in size; they are the largest of all the cetaceans. In fact, they are largest animals that have ever lived, exceeding in size even the largest dinosaurs. The leviathan of the group is the blue whale, reaching a length of over 100 feet and a weight of well over 100 tons. Even one of the smallest of the group, the minke whale, reaches a length of approximately 33 feet.

The Odontoceti include the sperm whale and the killer whale as well as the dolphin and the porpoise. And the latter two terms may need some clarification. There are morphological differences that make the dolphin distinct from the porpoise. Dolphins have a pronounced snout (the bottlenose dolphin has an extended upper and lower jaw that forms its snout from which it derives its name) and conical teeth. In contrast, porpoises have a small snout and flat, incisor-like teeth. (And to add to the confusion, there is a tropical fish often known as a dolphin.)

Seiners of the Sea

Along the eastern Pacific, there are three distinct families of mysticetes: Eschrichiidae (including the gray whale), Balaenidae (including the northern right whale), and Balaenopteridae (including the minke, sei, fin, humpback, and blue whales). Members of the Balaenopteridae family are commonly referred to as rorquals, taking their name from French *rorqual*, which comes from the Norwegian word *royrkval*, meaning "furrow whale." All rorquals have a series of longitudinal folds of skin (furrows) on their throats. These grooves are one of the distinguishing factors that set the balaenopterids apart from other whales.

One may wonder why there are so many distinct species of baleen whales and why each species is different from the others. Some species, for example, are like large container ships plowing through the water.

Others are sleek and streamlined. They range in color from midnight black to various shades of brown, gray, and blue. The fin whale, though, is asymmetrically colored: the right side of the head is white, and the corresponding area on the left matches the dark pigmentation of the body. Some have distinctive markings on fins, flukes, and backs, whereas others have callosities, raised thick patches of skin, on their heads. Some have long, spindly flippers; others have tiny, paddlelike ones that seem to date back to ancient times. The answer to this great diversity appears to be based on the variety of ecological niches in the ocean they exploit and differences in behavioral patterns. Simply, each whale has a unique relationship with its environment and developed morphologically to best adapt to its new home.

The key to understanding whales is to become knowledgeable about the foods they eat, even though most of their prey items may not be household names. Food governs every aspect of a whale's life and behavior from shape and size to distribution and foraging strategies. Most mysticetes feed primarily on tiny crustaceans and small schooling fish in shallow waters.

Although gray whales (*Eschrichtius robustus*) primarily feed on bottom organisms, their foraging behavior can be more complex, exhibiting a more generalist approach to the selection of food items. Resident gray whales near Vancouver Island were found to feed on a kind of amphipod (*Atylus borealis*) that swarms just above the bottom during the month of May. For the next month or so, they shift their attention to more abundant prey—mysids (shrimplike crustaceans). Exhausting that food supply, the whales feed on planktonic porcelain crab larvae that are near the surface. By August, the whales are feeding on bottom-dwelling *Ampelisca* amphipods, and they maintain a pattern of probing the bottom sediments for several months.

The right whale (*Eubalaena glacialis*) feeds upon copepods, and all minke, sei, fin, and blue whales feed on euphausiids—shrimplike zooplankton, often referred to as krill—to some extent. While the blue whale (*Balaenoptera musculus*) feeds almost exclusively on krill, the sei includes copepods in its diet, and the minke, humpback, and fin add fish.

The differences in diet among the rorquals, though all are similar looking, may reflect a body-size continuum. The minkes (*Balaenoptera acutorostrata*), the smallest in the group, are the most active, taking on fairly elusive prey in small patches because of their maneuverability.

Blue whales, the behemoths of group and with less mobility, concentrate on less elusive prey in larger patches. Finback (*Balaenoptera phycalus*) foraging behavior falls between that of the minkes and blues, and sei (*Balaenoptera borealis*) behavior between the minkes and fins.

The blue whale's large size does convey an advantage: enabling it to travel long distances as it seeks large concentrations of food. Though the much smaller minke does migrate, it is often content to stay in a specific area for a longer period because it can shift food species, and its body is not biologically engineered for long migrations.

The nature of an individual whale's baleen plates—the coarseness of the hairlike fibrous fringes, the density of fibers, the number of plates, and the length of the plates—varies markedly between species and is related to prey items captured by the filtering process. Gray whales, for example, have a very coarse filtering structure that allows them to separate benthic amphipods from the bottom sediments. Right whales, in contrast, possess a fine filtering mechanism, composed of more than 350 baleen plates that exceed ten feet in length, which is capable of capturing copepods less than a tenth of an inch in length.

Depending on their different baleen characteristics, the mysticetes exhibit different feeding strategies: sediment straining (gray whales), skimming (right and sei whales), and gulping (blue, fin, minke, and humpback whales). A gray whale bottom feeds, taking in huge quantities of food, water, and sediment in a gulping-sucking action while pressing one side of its mouth against the sea floor. (It usually feeds on its right side.) Water and mud are then expelled in a dark cloud, while benthic organisms are filtered out by the baleen. The whale's tongue dislodges the prey items from the baleen plates, allowing it to consume the food.

Skim feeders swim slowly with their mouths ajar, usually near the surface, sieving prey in their baleen plates as seawater flows through them. Skimming may last for several minutes or up to hours depending on the density of the food patches.

Gulp feeders swim with their mouths wide open; their flexible throat grooves stretch to increase the volume of water they can take in—as much as four times the amount possible without the grooves. The technique is akin to the way the pouch on the underside of a pelican's bill expands when it attempts to catch fish. The larger whales can gulp phenomenal amounts of water, as much as fifty to sixty tons, a weight

Humpback whale baleen (John Tunney/Shutterstock.com)

heavier than its own, in a few minutes. They are essentially enormous swimming mouths with tails attached. After gulping all that water, the tongue and elastic throat grooves force the water back out of the mouth through the baleen plates, which trap the food organisms.

A variation of the above feeding behavior is known as lunge feeding, in which a whale swims vertically to attack a concentration of prey instead of cruising along at a particular depth. It is explosive in nature, with the whale rapidly beating its tail fluke to accelerate and opening its mouth to about ninety degrees. This action generates the water pressure required to expand its mouth and engulf and filter a huge amount of water and fish. After the whale closes its mouth, the immense size of the engulfed water mass is evident from the whale's bloated shape in the head region. In less than a minute, all the engulfed water is filtered out of the distended throat pouch as it slowly deflates, leaving only the prey inside its mouth. This extreme lunge-feeding strategy is exhibited exclusively by rorquals and has been described by Paul Brodie of the Bedford Institute of Oceanography as the "greatest biomechanical action in the animal kingdom."

A new sensory organ found inside the jaws of some of these lunge feeders may have facilitated the evolution of the largest animals ever. This gel-filled organ, about the size of a basketball in a blue whale, appears to help blue and humpback whales coordinate the movements of their massive mouths. The organ senses vibration, pressure, and movement and subsequently sends this information to the brain to kick-start the lunge-feeding process.

To increase their feeding success, humpback whales (*Megaptera novaeangliae*) employ a unique technique known as bubble net fishing, which is analogous to humans setting a net around a school of fish. When coming upon a concentration of euphausiids, a single humpback whale may take a number of deep breaths before diving and seeming to disappear. But in a short time, a single air bubble rises to surface, followed by another, and soon a cloud of air bubbles streams up to the surface, forming a ten- to twenty-foot circle or bubble net. With the prey now densely packed together by the closed net, the humpback ascends, with mouth agape, through the center of the bubble net, capturing copious amounts of euphausiids. As long as these food items remain in the area, the humpback may feed in this manner for hours.

A more complex version of this feeding strategy involves the coordinated effort of six to a dozen or more humpbacks all releasing bubbles simultaneously to form a huge net, which encircles a school of small fish, such as herring. With the herring herded into a tightly packed bait ball, each whale will blast up through the surface and gulp concentrated mouthfuls of prey. This ability to work together is truly remarkable, considering the herring have the ability to escape.

At this stage, you may be wondering how these immense whales can ingest enough prey to meet their energy needs. The right whale, for example, which feeds solely on tiny copepods—roughly the size of a grain of rice—needs to consume between 400,000 and 4 million calories every day. To put those numbers in perspective, the daily caloric requirement of the right whale is equivalent to that of approximately 200 to 2,000 people. The answer to the question lies in finding and exploiting dense concentrations of prey. If the prey is widely dispersed, it may not be worthwhile from an energy standpoint for the whale to swim with its mouth open, straining the weak sea broth. Studies have shown

Bubble net feeding (Kent Ellington/Shutterstock.com)

that right whales feed on patches with a minimum concentration of 4,000 organisms per cubic meter of water. In a mostly barren sea, this prey density is generally confined to hydrographic fronts and upwelling regions, but the distribution of zooplankton in the water column may itself vary, necessitating different forage strategies on the part of the right whale. Flexibility in feeding is the key to obtaining a mouthful of food. When the copepods are near the surface, as they often are during upwelling events, the whale skims the surface, swimming with its upper jaw high above the surface. When the patches form a little deeper, the whale adjusts accordingly, feeding at depth and only surfacing every twenty-five minutes or so to breathe. In both cases, the mouth remains open except when the whale decides it is time to remove the food items from its baleen plates—about once an hour. If the copepods are found concentrated at even greater depths, such as in the deep scattering layer, it would not be long before the right whale descends to partake of this prey richness.

As with other creatures of the California Current Ecosystem, whales must be flexible in their behavior, adapting to changes occurring with the fluid environment in which they reside. The availability of food, or the lack of it, is a major impetus for the movement of whales along the Pacific coast. To the delight of the legions of whale watchers, mixed pods of whales—humpbacks, blues, and minkes—may stretch across the sea surface as far as one can see. To the casual observer, it is hard not to compare these migrating marine mammals with the vast mixed-species herds of zebras, wildebeests, antelopes, and giraffes that graze on the African plains.

The quintessential whale migration is that of the California gray whale, which travels from its wintering breeding grounds in the Gulf of California to its primary summer feeding haunts in the Bering, Chukchi, and Beaufort Seas—a passage of over 6,000 miles. The first account of gray whale migration may be from Charles M. Scammon, who wrote that gray whales follow "the shore so near that they often pass through the kelp near the beach." Though a whaler by necessity, Scammon was a naturalist at heart. His book, *The Marine Mammals of the Northwestern Coast of North America*, was not a commercial success at the time of its publication in 1874 but ultimately would be drawn upon extensively by generations of marine mammalogists. His work would eventually warrant election to the California Academy of Science.

The selection of the Gulf of California as a nursery ground for the gray whale may offer a major selective advantage: reducing the risk of killer whale predation on newborn calves in these low-latitude waters. Killer whale abundance is substantially greater at higher latitudes than at lower latitudes, and most killer whales do not seem to migrate with other whales, preferring to be near their primary marine mammal prey, pinnipeds. Following the gray whales south would remove the killer whale from the main pinniped population along the Pacific coast.

The gray whale is not the only epic traveler among the baleen whales. Humpback whales have been observed and identified off the coast of Japan in March and then seen in August of the same year near Vancouver Island. While the general populace delights in these herculean feats of migration of these giant creatures, long, arduous journeys are the exception rather than the rule among baleen whales. No barriers prevent whale movement; externally, there are no constraints. But for whatever reason—energy or social factors—travel distances can tend toward the conservative side. The blue whale, for example, travels from the coast of northern California to the Gulf of California in the winter—a journey considerably less than that of the gray whale.

How do whales, regardless of the length of their journey, arrive at specific destinations, then return to their departure points? How do they navigate in a relatively featureless environment while maintaining a well-defined social structure? How do they find and communicate the location of highly variable food sources during their travels?

The answers to these questions and other behavioral puzzles appear to be based on the acoustical world of the whales. Whales rely on acoustics to have a complete understanding of their environment—recognition and definition of places, friends, foes, and prey. Sounds, both those made by whales themselves and those coming from the environment, are key to the life activities of baleen whales.

In the often noisy undersea world, baleen whales use sounds in myriad ways but to different degrees. Grays apparently have a smaller vocal repertoire than either rights or humpbacks. The latter species, in particular, exhibits a sophisticated acoustic display, known informally as the "song." It has been observed that humpbacks sing most frequently during the breeding season. Researchers initially assumed that the singing by a male was the means of attracting a female, but some new observations have led to a different view: a display of dominance by a male, the acoustic equivalence of antlers or horns that show prowess and fitness. Blue and finback whales produce low-frequency calls, often below the range of human hearing, which under the right conditions can travel great distances—over thousands of miles—and can be used for navigation.

While our understanding of the acoustical world of baleen whales is still evolving, for decades we have been listening into the calls of these whales. Starting in the 1990s, collaboration between the academic community and the military sector yielded a startling result: sound recordings from an array of hydrophones positioned on the seafloor could be used to determine the large-scale movement and relative abundance of deep-water whales. These animals could be tracked and monitored for periods ranging from a few hours to many weeks, for distances from tens of miles to ocean-basin excursions.

Baleen whales have captivated our interest and curiosity, and for good reason. They can at times be both approachable and quite entertaining. We marvel at their antics, such as breaching, spyhopping, and spouting, as if they were putting on a show for us. And maybe they are, but perhaps a more nuanced view is in order.

The breaching of a multiton whale, as it launches itself out of the water, may be done for a variety of reasons, including the need to rid

Humpback whale breaching (Sue Leonard Photography/Shutterstock.com)

itself of parasites, to scare off predators, and, yes, just for the sheer joy of jumping out of the water. For a male, breaching may also be a signal of its power and strength to interested females. The crashing of multiton animal into the water sends a powerful sound wave over many miles.

Spyhopping involves the whale vertically poking its head out of the water, most likely to simply get a better view of the activity around it. What is it looking at? For a migrating gray, it might be to determine a coastal reference point. For a humpback, it might be as simple as looking back at the people on a whale-watching vessel. (And for a killer whale, it might be for locating prey.) A whale may spyhop for up to thirty seconds, during which it uses its pectoral flippers to keep it afloat, much like a person treading water using his or her arms.

The whale's spout or blow is not, as often told, a fountain of water spewing from the blowhole of the whale but a by-product of respiration. Whales can breathe only when they surface, and during that brief time, they have to take in and let out a lot of air quickly. Their lungs, the size of a small car, are strong enough to force out all of the "old" air in one breath. (Whales will exchange a far greater volume of air from their lungs, 85 to 90 percent, than humans, who only renew 10 to 20 percent of the air in their lungs.) Because the whale empties its lungs with such force, the air can travel more than thirty feet. The released air is warm and contains a copious amount of water vapor, which condenses upon entering the cooler environment above the sea surface, forming a stream of tiny water particles similar to when you "see your breath" on a cold winter day.

Though our perception of whales is often skewed by their playfulness and sometimes docile nature, we must keep in mind that they are still wild creatures, as they have repeatedly reminded us. In 1975, the late Kenneth Norris, acclaimed marine mammal researcher, experienced firsthand the ire of an adult female gray whale when he ventured into its calving waters in the lagoons off Baja California. The mother, angered that Norris and his group had restrained her calf for study, tried to sink their forty-five-foot vessel. Swimming directly under the boat, the irate whale periodically swung her flukes up in a determined attempt to disable the craft. While it was sheer luck that she did not sink the ship, the whale did indeed damage the propeller shaft when she repeatedly hit it with her broad tail.

The encounter with the mother whale may have shaken Norris, but his belief in the need to understand these creatures remained unshaken. Over a forty-year career, he would elevate the study and status of marine mammals to that of their terrestrial counterparts.

Toothed Whales

One of the best known of the odontocetes is the sperm whale. It is the largest of the toothed whale suborder, reaching lengths of over fifty feet and attaining weights of over forty tons. Immortalized in Melville's *Moby Dick*, it inspires emotions ranging from excitement to awe from those fortunate enough to come in contact with this majestic creature.

Along the Pacific coast, sperm whales make frequent appearances. In California waters, sperm whales are present year-round, but their numbers reach a peak from April through mid-June and from the end of August through mid-November. They are seen every season except winter in Washington and Oregon. While sperm whales are known to migrate, their movements are not as predictable or as well understood as that of baleen whales. From what we are able to decipher from the study of these reclusive creatures, males migrate to higher latitudes alone or in small groups and return to lower latitudes to breed. In contrast, females and their calves remain in tropical or subtropical waters all year long.

The sperm whale's scientific name, *Physter macrocephalus*, refers to the species's head, which typically makes up a quarter an individual's entire length. With its massive squared-off head, a comparatively small lower jaw, a highly developed trunk, and a scalloped tail, the sperm whale is unique in appearance among other cetaceans. How these features relate to the whale's life in the oceanic realm is still a mystery.

Typically, the sperm whale is dark bluish-gray to black in color, but on rare occasions white (albino) whales have been observed in nature. Although the great white whale in Melville's novel was fictitious, it, remarkably, was based on a real creature. Decades before Melville would begin to write his story, a huge albino sperm whale received widespread notoriety throughout the whaling community and even among the general public. The whale, Mocha Dick, had purportedly killed more than thirty men and had attacked numerous whaling vessels over the span of almost thirty years. A feared and cunning whale named for the island of Mocha off the coast of Chile, it most assuredly would have gotten the attention of Melville, who was familiar with the whaling profession, having sailed on the whaling ship *Acushnet* in 1841. If Mocha Dick was indeed the inspiration for Melville's own malicious creature, can we also conclude that Melville's description of an enraged Moby Dick attacking whalers and ships alike is correct? A watery demon bent on vengeance against his tormentors? According to present-day cetacean researchers, the historical view of the sperm whale as an evil monster is incorrect, one that most likely has been embellished over time by mariners and sailors because of the whale's large size and big teeth. What we know about these whales is that they are generally shy and easily startled by anything new in their environment. Even as far back as the nineteenth century, Thomas Beale, a whaling ship surgeon, published this description: "The sperm whale is a most timid and inoffensive animal . . . readily endeavoring to escape from the slightest thing which bears an unusual appearance." But can we summarily dismiss the eyewitness accounts by whalers of sperm whales exhibiting aggressive behavior? Were these rogue whales outliers from the norm?

Unfortunately, there is still a lot that we do not know about these creatures because we rarely get to study them firsthand, owing to their proclivity to remain submerged for extended periods of time. Because of their ability to efficiently store oxygen in their blood and muscles, adult whales can routinely dive to 3,000 feet or more and stay submerged for over an hour. One sperm whale, however, caught by a whaling ship in 10,000 feet of water, had a bottom-dwelling shark in its stomach, leading researchers to reevaluate the maximum depth a whale may reach.

Sperm whales spend much of their lives in the deep sea, feeding upon several species of abyssal squid. The squid range in size from the a few pounds to the swift and ferocious giant squid, which can be as long as an adult male sperm whale. Though the sperm whale has an impressive array of teeth in its lower jaw, they probably are not all that important in feeding. An analysis of the food items consumed by sperm whales has yielded an interesting fact: most were simply swallowed whole. Even a forty-foot-long giant squid weighing 440 pounds, recovered from a sperm whale, was completely intact, lacking any significant bite marks or punctures.

Giant squid, like all squid, have sharp beaks, which they use for biting into their prey. In the stomach of a large sperm whale, these beaks can accumulate, with numbers approaching the thousands. Over time, the pointed beaks irritate the whale's stomach lining. As a reaction to this irritation, the whale's intestines produce a cholesterol derivative commonly known as ambergris, which after being expelled by the whale ultimately hardens into a waxy aromatic substance. Ambergris has been known since ancient times when it would wash ashore, but to those who stumbled upon it, its origin remained a mystery. When heated, ambergris produced a pleasant earthy odor, which led the Greeks, Chinese, Japanese, and Arabs to hold it in high regard. But the real value of ambergris would ultimately be its use as a fixative in perfumes, which would, in part, contribute to the hunting of these whales during the eighteenth and nineteenth centuries.

Whalers may have been the first to impart the name sperm whale to this creature based on, well, pretty much as you'd guess, a white, viscous liquid found in the whale. While not the sperm of the whale, the liquid is oil, now known as spermaceti oil, which is produced by the spermaceti organ that fills most of the whale's huge head. Whalers soon came to realize the value of this oil as an exceptionally fine lubricant, but for quite some time, the role of the spermaceti organ remained a mystery.

After a number of wrong turns, including a belief that the organ plays a role in the buoyancy of the whale, it became apparent that the organ is important in echolocation: the production of a series of clicks that travel though the water, bounce off objects back to the whale, and are interpreted by the whale. In this way, they can judge distance, size, and shape as well as movement of a potential prey item. The spermaceti organ most likely focuses and amplifies the clicks made by the whale. Yankee whalers reported hearing clicks so regular and intense through their hull that they often referred to the sperm whale as the carpenter fish.

The spermaceti organ may serve other purposes as in male-male combat—a battering ram to injure and disable an opponent. Observations of aggression in this species suggest that head-butting during encounters of two males is a basal behavior of these cetaceans. A group of University of Utah researchers, led by David Carrier, proposed that the ability of a sperm whale to ram and destroy a stout wooden ship, many times more massive than itself, with its head stems from the primal act of aggression of one whale to another. To back up their argument, the researchers discuss the attack of the whaling ship *Essex* in 1821, which was the first documented case of a sperm whale deliberately ramming a vessel:

> It suddenly dived and surfaced less than 30 m from the ship traveling at estimated speed of 3 knots heading directly for the port side of the ship. The whale struck the ship, which shook "as if she had struck a rock" (Chase, 1821). The whale then swam approximately 500 m leeward from the ship, where it acted as if it were "distracted with rage and fury." After several minutes of this display, it swam directly in front of the ship and then charged the ship again, this time with a speed near 6 knots. The whale struck the *Essex* directly beneath the cathead and completely stove in her bows. The *Essex* started sinking, and capsized within 10 min on its port side.

To the researchers, a charging sperm whale has enough momentum to seriously injure a foe as well as to disable a stout wooden vessel of the whaling era.

Puzzles remain and questions are still unanswered about the sperm whale; nonetheless, a picture of this cetacean is coming into focus: one depicting a species that has over time exhibited great success in colonizing the ocean and exploring its deeper waters. But this success begets another question: which features of sperm whale society are responsible or have contributed to the mastery of their watery realm?

Sperm whales often travel in groups, known as pods, of some fifteen to twenty individuals. These pods are made up of females and their young, while males may roam solo or move from group to group. Within the pod, communal child care appears to be a common practice, which seems to start soon after birth when newborn calves are immediately tended to by other members of the pod. This babysitting is important because female sperm whales have only one calf every four years, making each offspring enormously valuable to the mother. Although large in size and intimidating in appearance, sperm whales do have potential enemies, no more so than the killer whales, which have been observed attacking sperms. To protect a calf, females may form a tight group, almost touching one another, while the calf is safely positioned in the center. Any attack by the killer whale is met by the group wheeling this way and that to meet their antagonist head on. Communal vigilance and cooperation is paramount to the survival of the defenseless calf.

Sperm whales are known to aid one another in other ways too. Females in a group may suckle one another's young, and females were observed tending to their harpooned companion, trying to break the line of the embedded harpoon.

These actions highlight not only the intelligence of these animals—they have the largest brain of any creature known to have lived on Earth—but also their ability to cooperate and understand the value of teamwork. We are finding out that sperm whale society is very complex, possibly in league with the better-known terrestrial counterparts—that of chimpanzees and elephants, which are recognized as being among the world's most cognitively advanced animals.

While cooperation among sperm whales was most likely necessary in the context of communal care for and protection of young, how might such a caring society come about? Evolutionary biologists believe that there are two possible sets of conditions that may foster the development of this unique interaction among members. The first is reciprocal altruism, which occurs when favors are interchanged. It requires that individuals within a group recognize one another, become familiar with one another, and spend considerable time together, so that many opportunities can arise to return favors. With regard to the first requirement, there is no reason to believe that these intelligent creatures cannot identify and distinguish among individual members. And because the social units are relatively permanent, made up of the same female members over the years, there is ample time to interact and cooperate in a positive fashion. The other circumstance that may foster cooperation is when the animals are related. Genetic analysis of whale skin has shown that the DNA markers are more similar among whales from the same groups than among whales from different groups. The pods appear to be made up of matrilineal family units—females spending most of their lives with their mothers and other closely related females—within which cooperation would be highly expected.

Group cooperation, it appears, may have been the dominating factor, if not the only factor (possibly complex brain, sophisticated echolocation), in the evolved ecological dominance of the sperm whale of its deep-water habitat. But the sperm whale is not alone in its preference for foraging at depth; elephant seals and swordfish routinely descend to feed on squid too, as do the ten-foot-long pygmy sperm whale and the smaller (eight-foot-long) dwarf sperm whale.

Beaked whales (family Ziphiidae), however, may be strongest rivals of sperm whales. Like sperm whales, they are toothed whales, but the two groups of deep-water whales are not closely related. Beaked whales, which derive their name from their elongated snout (they are sometimes referred to as bottlenose whales), are the least known of all the cetaceans. Due to their preference for deep water habitat, they are rarely seen; our study and knowledge of this species is still in its infancy. What information researchers have been able to gather about these reclusive whales comes from autopsies on stranded whales or those killed by whalers.

Though beaked whales comprise almost one-quarter of the eighty-seven extant cetacean species, they are rather rare in California waters. Beaked whale species known to occur in the California Current marine ecosystem include the Baird's, Cuvier's, Hubbs's, Blainville's, Perrin's, Stejneger's, ginkgo-toothed, and pygmy beaked whales. Though similar in body shape and coloration, the animals vary in size from about twelve feet for the adult pygmy beaked whale to forty-two feet for Baird's beaked whale. In the wild, beaked whales are hard to distinguish from one another, leading cetologists to speculate that possibly other species of beaked whales still unidentified and unclassified roam the ocean. It also appears that beaked whales specialize on only small parts of the deep ocean environment.

Satellite tags attached to Cuvier's beaked whales (*Ziphius cavirostris*) off the coast of California recorded a dive to almost 9,300 feet and one that lasted 137 minutes, bestowing on this marine mammal the title

Diving Depths of Selected Marine Organisms and Ocean Pressure

Species	Diving Depths (feet)	Pressure (lbs/in²)
Cormorant	150	70
Bottlenose dolphin	900	400
Jumbo squid	2,500	1,100
Leatherback turtle	3,300	1,500
Great white shark	3,500	1,590
Swordfish	5,500	2,500
Elephant seal	7,000	3,180
Sperm whale	7,400	3,370
Cuvier's beaked whale	9,300	4,230

of deep-dive champion. The feat exceeds that for either the elephant seal or sperm whale, also known for their breath-holding capabilities. In part, the key to the success of these whales is the very high level of the protein myoglobin found in their muscle tissues, which is a storehouse of oxygen. In humans, muscle tissues appear red due to this protein, but because myoglobin occurs in such high concentration in these whales, the tissues look almost black. But another key adaptation, probably unique to beaked whales, is the dramatic reduction in air spaces in their bodies. The decrease in air volume allows these whales to better withstand the crushing pressure at the great depths and also likely serves to reduce the uptake of dissolved gases into their tissues, which could potentially lead to decompression sickness.

The new depth mark set by the Cuvier's beaked whale is even more remarkable, and puzzling to researchers, considering the relatively small size of this species. As a general rule, the depth and duration of a dive tend to scale with body size and mass, but adult Cuvier's beaked whales are much smaller (5,500 pounds) than other masters of the abyss, such as the sperm whale (125,000 pounds). Though more data are needed to resolve this mystery, tagging these incredibly shy creatures is extremely difficult and depends not only on the perseverance and skill of the researcher but also a great deal of luck.

The lack of data on marine mammals in general is particularly disturbing. The International Union for the Conservation of Nature (IUCN) identifies 25 percent of marine mammals at risk of extinction, but the conservation status of nearly 40 percent of these animals is unknown.

But the numbers get worse. For beaked whales, 90 percent are data deficient. Population trends, in particular, for all beaked whale species are listed as unknown on the IUCN Red List.

To alleviate this data paucity on beaked whales, the National Oceanic and Atmospheric Administration's Southwest Fisheries Science Center in La Jolla, California, has been conducting cetacean abundance studies since 1991 by means of shipboard line-transect surveys, analysis of whale strandings, and models. The result of this multidisciplinary approach was not encouraging: a decline in abundance of beaked whales in the California Current Ecosystem during the study period. Causes for the decline may prove more elusive than the whales themselves, but at this time, three hypotheses have been put forth: incidental catch by fishermen, impacts of anthropogenic noise, and changes in the ecosystem.

While bycatch mortality of beaked whales has occurred in high-seas drift net fisheries, the regulations placed on the California drift gill net fishery—use of pingers on nets and large time-area closures in Central California—have most likely had the positive effect of reducing the potential fishery interactions with beaked whales in the California Current.

Ambient noise off the coast of California has increased significantly over the past several decades, and of primary concern for beaked whales is noise created by routine military sonar operations. Some species of beaked whales, including Cuvier's beaked whale, have experienced a gamut of responses to sonar, ranging from changes in behavior to death from strandings. In the Southern California Bight, beaked whales are exposed to sonar activities in the vicinity of the U.S. Navy's Southern California Anti-Submarine Warfare Range (SOAR). If sonar activities at SOAR are impacting beaked whales, one should expect the effects to be most prevalent in the vicinity of SOAR. And yet the aforementioned study of deep-diving Cuvier's whales was conducted in this very area, hinting that the whales may have become habituated to sonar. In addition, the researchers were shocked to find as many Cuvier's beaked whales as they did, considering that they seem to be sensitive to this type of disturbance elsewhere. The conclusion: the evidence to implicate noise from either naval operations or other sources as a cause of the possible decline of beaked whales is, at best, equivocal.

The California Current System is a highly variable and complex system, subject to interannual and interdecadal climatic events (El Niño,

Pacific Decadal Oscillation) that are associated with marked changes in the biological regime. Though the feeding ecology of beaked whales is poorly understood, the impact on prey items of beaked whales by changes in the California Current System cannot be easily dismissed. Mesopelagic (approximately 660 to 3,300 feet) fish biomass in the Southern California Bight has declined by as much as 60 percent since the 1980s because of increasing deep-water hypoxia.

Changes in an ecosystem can also occur when community structures within the system change, resulting in marked changes in food web dynamics. Since the 1990s, the Humboldt squid, *Dosidicus gigas*, has substantially expanded its geographic range in the eastern North Pacific by invading the waters off central California. While *Dosidicus* may be associated with warm-water temperatures at the center of its tropical distribution, it is a physiologically highly adaptable species, developing a tolerance for the cool temperatures found at depth. A large, aggressive predator, *Dosidicus* has been linked to the decline of cephalopods, a food item of beaked whales.

Recognizing the interactive effect of multiple changes in the California Current System is a challenge to those trying to understand population trends and to those concerned in ocean conservation.

Home of the Dolphin

Bring up the topic of "dolphin," and most people conjure up the image of the animal they have seen in aquariums worldwide and made famous by the *Flipper* television series: one with a bottlenose snout, the "smiley" face, and the friendly and playful demeanor. But there are many other dolphins (over thirty species in the family Delphinidae), including the largest true dolphin, the killer whale. All are streamlined, with small front limbs for maneuvering and broad tails for propulsion. But, as might be expected, members of the large Delphinidae family exhibit variations on this basic theme. Male killer whales, for example, have large rounded front flippers and a huge dorsal fin. Some dolphins have pronounced snouts; some have blunt heads without a trace of a snout. Some have a few teeth; others have well over a hundred.

Some dolphin species, such as the common dolphin (*Delphinus delphis*), roam the California Current in large pods. The number in a pod can be several hundred, but it can also be in the thousands. These marine animals bring to mind the great grazing herds of the African plains; all activities—feeding, detection of predators, mating, and caring for offspring—occur in this large group setting. The need for social interaction is the norm, not the exception, among dolphins.

Touch and emission of a variety of sounds are big parts of their connections with one another. While a lone dolphin must be totally vigilant at all times of its main adversaries—sharks and killer whales—in a coordinated society, only one member need detect a potential predator and sound the alarm by emitting a cacophony of sounds, thereby triggering the appropriate response, flight or fight, from the others within the group. (Dolphins create their sounds, such as whistles, in a closed system that is essentially airtight and transmit these signals outward form organs in their heads known as melons.)

Bottlenose dolphins (*Tursiops truncatus*) are particularly adept at coordinated hunting, each filling a specialized role during the hunt. Prey are first located by means of echolocation. A series of high-pulsed clicks transmitted and interpreted by the dolphins keys them onto their prey. In the group, which usually consists of three to six members, one dolphin acts as the "driver" and herds the fish into a tight ball toward the "barrier dolphins," who are tightly grouped together to prevent the fish from escaping. The driver then uses its broad tail to slap the water, creating quite a commotion and causing the startled fish to leap into the air. In this vulnerable position, the fish are easy prey for the dolphin team, lunging out of the water and seizing their bounty in the air. Some researchers have argued persuasively that individual foraging specializations, if socially learned, are intrinsic to cetacean culture, but it remains unclear why a division of labor with role specialization is so rare in terrestrial species that hunt cooperatively. It may simply be that in the terrestrial regime the practice does not warrant the effort and energy expenditure because it rarely improves performance. But other hypotheses center on the differences between the terrestrial and marine habitats, particularly with regard to prey diversity, biomass, and seasonality. Do dolphins have to be smarter than other organisms to survive in an environment where successful foraging can be a challenge?

As long ago as the seventeenth century, biologist and naturalist John Ray determined that dolphins had very large brains for their bodies, with a complexity, at least to him, similar to that of humans. Assessing and defining intelligence, whether in apes or dolphins or even humans, is a slippery endeavor, fraught with biases and uncertainties. But, in fact, some dolphins are quite *intelligent*. The bottlenose dolphin, for example, has a brain-to-body ratio, by weight, approaching that of humans. It also possesses large amounts of gray matter, the brain tissue largely responsible for high-level cognitive functions. In addition, we know that the species is highly adaptive and capable of learning complicated tasks while in captivity.

Along the Pacific coast, the common dolphin is one of the most abundant dolphin species, ranging from Ecuador to British Columbia, but most abundant north of Monterey Bay. In these waters, two separate species exist: a short-beaked form that resides in deep waters and a long-beaked one that occurs primarily in shallower waters. The two species can be difficult to tell apart—differing slightly in physical size (six to eight feet in length), coloration (gray body and white belly), and pattern. (It was not until the 1990s that genetic analysis proved the existences of two distinct species.)

The common dolphin is not a deep diver as are sperm and beaked whales. The short-beaked species, in particular, feeds primarily at night—resting during the day—when the deep scattering layer of the California Current Ecosystem rises to shallower depths. Stomach analyses of these dolphins show that they mainly consume a variety of deepwater fish and squid. Long-beaked common dolphins, in contrast, consume a variety of small schooling fish, including anchovies, sardines, and pilchards.

The Pacific pilot whales (*Globicephala macrorhynchus*), one of the largest of the oceanic dolphins, exceeded in size by only the killer whale, are easily distinguishable from other dolphins by their distinctive large bulbous melons. The derivation of the name is from the Latin *globus*, for "globe or ball"; the Greek *kephale*, for "head"; and the Greek *macro*, meaning "enlarged," and *rhynchus*, meaning "snout or beak." The common name "pilot whale" lies in the belief that there is a leader or "pilot" of the pod. To add to the list of names (and maybe confusion), they are sometimes known as blackfish, most likely due to their overall dark gray or black coloration.

Pilot whales are excellent divers, capable of diving to depths of almost 3,000 feet to pursue deep-water squid. But compared with sperm and beaked whales, foraging pilot whales are more energetic at depth, their dives characterized by bursts of speed. Deep sprints appear to be in opposition to the accepted expectation that deep-diving mammals will swim at relatively slow speeds to conserve oxygen and maximize foraging time at depth. But pilot whales may have developed this tactic to

Pod of orcas (Karoline Cullen/Shutterstock.com)

exploit a previously untapped deep-water niche characterized by high-caloric but fast-moving prey. This energetic foraging tactic, focused primarily on a single prey item, draws a comparison to the high-risk but high-gain strategy of some terrestrial predators, such as cheetahs, that employ explosive bursts of speed to chase down their fleeing prey.

Of all the cetaceans, pilot whales are most likely to beach themselves. A key factor in these mass strandings appears to be the strong social cohesion found within pods of pilot whales. A "follow me" type of syndrome may take place when the pod follows a member of high ranking within the group. This lead whale may have become disoriented or exhibited signs of sickness or old age.

What's in a name? The killer whale is also referred to as orca whale or orca and less commonly as the blackfish. The English-speaking populace most often uses the term "killer whale," possibly due to the impression, at least from the human perspective, of the animal being a savage and bloodthirsty hunter. Pliny the Elder may have been the first to form this image, writing that "orcas (the appearance of which no image can express, other than an enormous mass of savage flesh with teeth) are the enemy of [other whales]. . . . They charge and pierce them like warships ramming." True, it is an apex predator, lacking any natural enemies, but our view may be skewed. No doubt the mackerel or the squid regards the playful common dolphin as an equally vicious predator. And yet proponents of the "killer whale" name argue that it has a long heritage, pointing out that the genus name *Orcinus* means "of the kingdom of the dead." Since the 1960s, the use of the term "orca" has grown in popularity to downplay the "killer" stereotype and because, being part of the family Delphinidae, the species is more closely related to other dolphins than to whales. Regardless of the name, this marine mammal routinely preys upon other marine mammals. In addition to pinnipeds, thirty-two cetacean species have been identified, from examining stomach contents, as killer whale prey.

As with other dolphins, killer whales are highly social—swimming, feeding, and hunting in groups. But research conducted during the 1970s and 1980s off the west coast of Canada and the United States identified three distinct pods.

- Resident: These are the most commonly sighted and intensively studied of the three pods. Their home range is in the coastal waters off Washington and British Columbia, where they mainly feed on fish. Salmon, in particular large, fatty Chinook and chum, account for approximately 95 percent of their diet. The smaller sockeye and pink salmon are not a major prey item. These whales have a particularly complex and stable social organization. As opposed to other mammal social structures, the offspring will live with their mothers for their entire lives. Because killer whales can live up to ninety years, as many as four generations travel together, separating only for a few hours to feed or mate. These matrilineal family groups are the most stable of any animal species.
- Transient: These whales roam, generally in small groups of two to six animals, widely along the Pacific coast, having been sighted as far north as Alaska and also in California waters. When their range overlaps that of resident killer whales, they will assiduously avoid making contact with the residents. The name "transient" appears to have originated from the widely held belief that these killer whales were outcasts from larger resident pods. But new research has revealed that transients are not born into resident pods or vice versa. The split between the two groups may have occurred as far back as 2 million years ago, and genetic analysis indicates that interbreeding has not occurred for up to 10,000 years. Family bonds within transient pods are not as strong as those found in residents, with extended or permanent leaves of the offspring from the group.

- Offshore: This is a relatively new group, having been discovered in 1988. As the name implies, they travel far from shore and are most commonly found off Vancouver Island and near the Queen Charlotte Islands in British Columbia. Little is known about their habits and behavior, but what we do know is that they are genetically distinct from residents and transients.

Like other dolphins, killer whales rely heavily on acoustics to communicate, find prey, and navigate. Resident groups in the Pacific tend to be quite vocal, emitting a variety of sounds, including clicks, whistles, and pulsed calls. The clicks are believed to be used primarily for locating and discriminating their prey. In contrast, transients are relatively silent when hunting since marine mammal prey hear well underwater at the frequencies emitted by the whales. They emit a single click but avoid the long train of clicks observed in other populations.

Whaling

For almost as long as humans have taken to the sea, whales have been hunted. Living along what is now the northwest coast of Washington, the Makah Tribe hunted whales at least 2,000 years ago. Makah elders say they have hunted whales "forever," so central were whales and whaling to their culture. For the Makah, whaling benefited the entire community; whales may have provided up to 80 percent of the sustenance needs of tribal families.

The preparation for the hunt was an elaborate affair marked by a number of rituals, such as praying, fasting, and bathing in icy streams, by each whaler. Each man readied himself spiritually and physically for the arduous undertaking upon which he was about to embark.

When the time approached, generally marked by a period of fair weather and calm seas, the whalers would paddle out in their canoes to intercept a migrating whale, usually either a humpback or gray. Waiting patiently in the cold waters of the Pacific, they knew from experience what to expect from the whale. Sooner or later, it must surface to breathe, an opportunity not to be lost, when a harpoon could be thrust into the flank of the beast. Attached to the harpoon's line were several sealskin air bladders, which were used to retard the wounded whale from diving. Upon the death of the whale, it was towed back to shore, aided by a tribe member who dived into the icy water to lace the mouth of whale shut to prevent it from filling with water and sinking, where the carcass would be butchered. The whale meat and blubber would be divided up among the villagers according to a strict tribal hierarchy. Humpbacks were of more value to the Makah than grays because of their tasty meat, and subsequently most of the whale would be eaten. The less palatable gray whale would be rendered for oil. This time-honored tradition of whale hunting had little impact on the whale population. But that would change. Where sustainability and conservation were at the forefront in Native American society of the Pacific coast, the actions of foreign nations would be dominated by greed and overexploitation.

In 1778, Captain James Cook, indefatigable Pacific explorer, sailed into Nootka Sound on the northwest coast of America. Though the main thrust of his voyage had been to find the infamous Northwest Passage, which would link England to the rich markets of the Orient, he became enthralled with the relatively sophisticated whaling operations of the local natives. Though he arrived too late to observe the migration of humpback and gray whales, it surely must have occurred to Cook that commercial whaling, similar to that already in place in Greenland, would be a suitable and profitable undertaking, ultimately propelling British expansion into the North Pacific Ocean. Soon, where that had once been only fur-trading vessels and the occasional reconnaissance ship, there were now whale ships. And over the decades, the numbers would increase markedly, which when coupled with improved technology would lead to the depletion of each whale species. Gray whales, for example, were hunted extensively throughout the nineteenth century, leading them to the brink of extinction. The lagoons of Baja California were the primary killing grounds of gray whales. In 1877, Charles Melville Scammon wrote that "the large bays and lagoons, where these animals once congregated, brought forth and matured their young, are already nearly deserted. The mammoth bones of the California gray whale lie bleaching on the shores of these silvery waters, and are scattered among the broken coasts from Siberia to the Gulf of California; and ere long it may be questioned whether this mammal will not be numbered among the extinct species of the Pacific."

By the mid-nineteenth century, North Pacific right whales had experienced a marked decline in numbers due to uncontrolled hunting; the humpback whale population was similarly reduced prior to World War I; and finally, the fins and blues were decimated—a tragic story of humankind's greed and wanton destruction of these creatures.

Even with "improved" technology, the hunting was often grossly inefficient. Many whales—probably 50 to 80 percent—simply perished on the ocean floor, as the whalers were inept at bringing them back to shore. While it would be naïve to believe that the Makah landed all their harpooned whales, we might conclude that these European whalers had not absorbed, or simply ignored, the tried-and-true techniques of the native hunters.

Commercial whaling also left a significant cultural legacy in this part of the world. The industry was a major player in crucial transformation from a mainly subsidence society to one hell-bent on the unlimited exploitation of a resource to fuel the growing economic engine of the Pacific Northwest. The difference between Native and European whaling came down to technology. With advanced whaling practices on their side, Europeans could more efficiently pursue whales for longer periods and over greater distances. The result: by the mid-twentieth century, whales had become commercially extinct.

In 1946, alarmed by the precipitous decline of whales around the world, a number of countries established the International Whaling Commission (IWC), which was charged with the conservation of whales and whaling. In 1986, the IWC introduced zero catch limits for commercial whaling, a provision that is still in place today. In the United States, whales are protected by Marine Mammal Protection Act (1972), and endangered or threatened cetaceans are further protected under the Endangered Species Act (1973).

Some gains have been made. Gray whales have mounted a remarkable comeback. From a low of 4,000 to 5,000 individuals in the sixties, the population has grown to as many as 25,000, close to its historic high. In 1994, the gray whale was removed from the endangered species list. Even the humpback whale population, once perilously low, has steadily increased, albeit slowly, and is now being considered recovered. Constant vigilance is needed if we, as a society, are to protect cetaceans from a growing number of human threats (ship collisions and pollution) as well as from climatic changes that impact other marine mammals.

Chapter Seven

LOOKING FOR SOLUTIONS
Protecting the California Current Ecosystem

Almost since humans first set sail, they have long made claims on the sea. The fifteenth and sixteenth centuries would usher in a period of expanding maritime dominance by Spain and Portugal, which would conspire to partition the Atlantic Ocean between them. Spain would have domain over the waters of the western Atlantic and the Gulf of Mexico, and Portugal would have exclusive rights to the remainder of the Atlantic. The boundary separating these outlandish claims was arbitrarily set at 35° west longitude based on a papal decree from Pope Alexander VI. A dark period of closed seas, or mare clausum, would descend on the rest of Europe.

While these claims were mainly implemented to facilitate Spain's and Portugal's quest for the riches found on land, this mindset of staking title would in the future extend to the biological resources of the seas. Stemming from our own hubris and naïveté, we would apply the concept of boundaries to the oceans' creatures.

Since humans have generally been ignorant of the natural linkages that exist within and across marine ecosystems or of the propensity for some organisms to undergo prodigious migrations, they have arbitrarily divided the oceans' waters for management purposes. Centuries ago, coastal nations agreed in principle that their individual maritime claims extend a fixed distance from the land. An acceptable boundary was implemented in 1702 of three miles—the distance of a cannon shot. Since the mid-twentieth century, numerous nations have come to view the three-mile limit as an archaic concept of the international law of the seas and have been eager to extend their reach. By 1982, territorial claims were creeping seaward when the United Nations Convention on the Law of the Sea defined territorial seas as a zone extending twelve nautical miles from the mean low-water mark of a coastal state seaward, over which the state has sovereign rights. In addition, the Law of the Sea treaty guaranteed the Exclusive Economic Zone, which extends offshore for 200 nautical miles and in which a coastal nation has complete control of economic resources, including fisheries. While these zones are deemed adequate to assign jurisdiction, they do little to promote the effective management and governance of marine ecosystems. All these changes were nothing more than political posturing, an attempt to limit foreign fleets from competing with domestic fisheries for a limited resource. The result: commercial anglers overexploited the fishery. Some, such as the North Atlantic cod fishery, have not rebounded to population levels prior to intensive fishing.

Marine management is not simple. The challenge for protecting the marine environment goes well beyond the extension of national and international laws. Marine ecosystems, in particular, represent a rich and diverse group of coevolved species that have complex and nonlinear interactions in a shared environment. This has made them difficult to manage, and the recent record of exploitation suggests that previous management schemes were failures. Over the last few decades, there has been a shift in management circles away from the traditional single species or piecemeal approach to a more comprehensive-based method. This tack requires managers to consider all aspects, including habitat, predation, forage, competition, and meteorological-oceanographic processes, of the environment that might affect the population of a particular species. It is simply not possible to study changes in a single species without viewing the compensatory changes in the ecosystem or modification of the physical/chemical components of its habitat.

Past attempts to manage marine ecosystems have generally fallen short because, as you might guess, scientists and managers really do not understand how to go about such a complicated and all-inclusive scheme. The task is even tougher for the California Current Ecosystem because it is made up of several overlapping smaller ecosystems—nearshore and offshore, pelagic and benthic—with widely ranging conditions from north to south. Ecological assessments are hindered by significant variability inherent in the system. As some have characterized the California Current Ecosystem, it does not ever exist, really, in an average state.

While recognizing the difficulty of managing large swaths of the ocean, a groundswell was occurring during the late 1960s in the U.S. government to protect scenic coastlines and special marine places. The impetus, at this time, was the growing concern of the rapid expansion of petroleum operations into offshore waters. But what really caught the attention of the public was the 1969 oil spill that occurred in the Santa Barbara Channel due to an oil rig blowout. At that time, it was the largest oil spill in the United States, and the sight of oil-soaked seabirds and dead marine mammals generated intense press coverage, public outrage, and, ultimately, government response.

National Marine Sanctuaries

In the early 1970s, the environmental movement was sweeping across the United States, and the topic of ocean degradation was on the minds of many people. In 1972, the floodgates of environmental legislation opened, with Congress passing a number of environmental laws, among them the Marine Protection, Research and Sanctuaries Act. The act was ambitious in its scope, authorizing a trio of programs to protect and restore marine ecosystems. Proponents of the legislation argued that new regulations, such as controlling the dumping of wastes in ocean, coupled with initiatives to study the long-term effects of humans on ocean ecosystems, would be a major sea change in how we manage critical habitats. The Secretary of Commerce would have the authority to designate national marine sanctuaries for the purpose of restoring or protecting the natural and cultural resources of marine areas. Only human uses deemed compatible with this goal would be allowed. The proposed sanctuaries would take as their model the system of wilderness areas on land that were established by the Wilderness Act of 1964.

Unfortunately, the Sanctuaries Act fell short of following in the footsteps of its terrestrial counterpart due primarily to interpretation of intent of the act itself. While the Secretary of Commerce would establish sanctuaries for preservation and restoration purposes, Congress would allow for both preservation and extractive uses in the sanctuaries. But what exactly constituted an extractive use? Would drilling for oil and natural gas be allowed? Did fishing fall under this heading? And, if so, what type—commercial or recreational? No answers were provided to these questions, and over time, Congress simply gave the Secretary of Commerce the discretion to decide which uses in a particular sanctuary would be compatible with the resource protection objectives of the act.

Not everyone was thrilled about the idea of vast swaths of the ocean being designated as sanctuaries. Oil and commercial fishing industries, as might be expected, were increasingly antagonistic toward the program because of its potential to limit or prohibit their activities. The oil industry petitioned to have exploration and development activities continue unabated in sanctuaries, and the fishing sector was opposed to any access restrictions to fishing grounds that it had historically used. From roughly 1977 until 1986, the fishing and petroleum interests, along with their congressional allies, challenged the Sanctuaries Act. Battles raged over individual sanctuary proposals and helped stall any significant implementation of protective measures. Although the oil and fishing industries failed in their attempt for an outright repeal of the act, they were largely successful in limiting its scope. In spite of the reservations of some scientists and managers about the effectiveness of the proposed sanctuaries and the ambiguities within the Sanctuaries Act, Congress set the wheels in motion to designate specific areas needed for protection.

In 1977, California nominated the Channel Islands, Gulf of the Farallones, and Monterey Bay as potential marine sanctuaries. In 1980, the Channel Islands became the first marine sanctuary along the Pacific coast, to be followed the next year by the waters off the Farallones.

The Channel Islands National Marine Sanctuary, a site of exceptional biodiversity, from microscopic plankton to mammoth whales, has an area of 1,470 square miles and encompasses the waters that surround the Anacapa, Santa Cruz, San Miguel, Santa Rosa, and Santa Barbara Islands (five of the eight Channel Islands of Southern California). The boundaries of the sanctuary extend from the mean high tide to six nautical miles offshore around each of the five islands. Most of the sanctuary (almost 92 percent) is comprised of deep-water habitat, with depths ranging from 100 feet to over 5,000 feet.

Oceanographic conditions in the sanctuary are profoundly influenced by the complex geography of Point Conception, a prominent promontory north of the sanctuary. The orientation of the coastline abruptly shifts from north-south to east-west at Point Conception. During the spring-summer upwelling period, nutrient-enriched waters bathe the western islands of San Miguel and Santa Rosa, accounting for high biological productivity. In contrast, the waters of the eastern islands of Anacapa and Santa Cruz are warm and are home to subtropical species, which have been transported northward by the California Countercurrent from Baja California.

The Gulf of the Farallones National Marine Sanctuary spans 1,282 square miles and lies to the west and north of San Francisco. It includes

Anacapa Island, Channel Islands (Joseph Sohn/Shutterstock.com)

nearshore waters up to the mean high tideline from Bodega Head to Rocky Point in Marin County and offshore waters extending beyond the Farallon Islands and the continental shelf.

The sanctuary is impacted by the strong upwelling that occurs near Point Reyes. The upwelling filament that extends from the promontory bathes the sanctuary with nutrient-rich water, supporting a vibrant and diverse assemblage of organisms that managers realized needed protection.

In 1983, the National Oceanic and Atmospheric Administration (NOAA), an agency within the Department of Commerce and charged with evaluating nominations for sanctuaries, removed Monterey Bay from its list of active candidates, arguing that similar marine environments were already protected by California's relatively two new sanctuaries. But the people of California were steadfast in their opinion that Monterey Bay was biologically unique and should be placed under the sanctuary umbrella of protection. After years of grassroots action and negotiation, Monterey Bay was designated a sanctuary in 1992. With its inclusion into the network of sanctuaries, it became the largest marine sanctuary, encompassing 5,322 square miles of ocean, which is larger than the state of Connecticut. In 2009, the sanctuary was expanded by 775 square miles to include the Davidson Seamount, one of the largest underwater mountains (23 nautical miles long and 7 nautical miles wide) in U.S. coastal waters and home to a variety of marine species. This undersea mountain, located 70 nautical miles west of Monterey, has been referred to as "oasis in the deep," hosting large coral outcrops, vast sponge fields, deep-sea fish, and a high number of rare benthic species—many of which have not been fully identified.

While the fight to recognize Monterey Bay as a sanctuary site wore on over the years, in 1989 federal officials designated Cordell Bank as a marine sanctuary. The Cordell Bank National Marine Sanctuary is entirely offshore, with the eastern boundary six miles from shore and the western boundary thirty miles from shore. It is the smallest of the sanctuaries at 529 square miles.

The Cordell Bank, Gulf of the Farallones, and Monterey Bay National Sanctuaries have contiguous boundaries. While each has distinct features and settings, many resources are similar and some even move freely between the sanctuaries; therefore, individual sanctuary management is not always determined by site boundaries. To protect the natural and cultural resources, staff members of the three sanctuaries share responsibilities for research, monitoring, education, and management plan development. (In early 2015, the Obama administration approved the expansion of the Cordell Bank and Gulf of the Farallones Marine Sanctuaries, more than doubling their size.)

In 1994, the last of the Pacific coast marine sanctuaries came online—the Olympic Coast National Marine Sanctuary (OCNMS). Located off the Olympic coast of Washington, it encompasses 3,189 square miles of Pacific Ocean and stretches from Cape Flattery in the north to the mouth of the Copalis River. Extending twenty-five to forty miles from the coast, the OCNMS includes most of the continental shelf, as well as segments of three submarine canyons—the Nitinat, the Quinault, and the Juan de Fuca.

Upwelling tends to be more sporadic and weaker in the sanctuary than farther to the south owing to unfavorable winds. Nonetheless, when upwelling does occur, nutrient-laden water is drawn up from the deep-water canyons. The result is a highly diverse food web, ranging from many species of kelp to a menagerie of foraging marine mammals.

Though there are five Pacific sanctuaries, encompassing at present a total area of almost 20,000 square miles, there are geographic "holes"

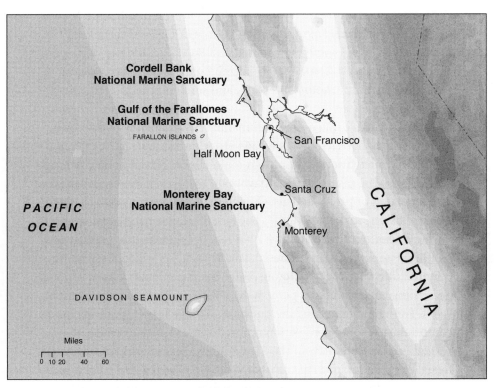

Marine Sanctuaries

in the system, with no protective areas in Oregon and Alaska. And since Congress enacted a moratorium on new designations in 2000, there appears to be little hope for total protection of the vast California Current Ecosystem in the near future.

Have the existing sanctuaries been a success? The answer to that question depends on who you ask and how you interpret success. Some have argued that the sanctuaries provide refuges for threatened seabirds and pinnipeds; others have trumpeted the banning of oil and gas drilling in these sensitive areas. The sanctuaries have also served as focal points for educating the public about marine conservation and as natural laboratories for study of Pacific wildlife. But although the sanctuaries do offer some protection to nationally significant marine areas, they come up short as a complete preservation system.

The Sanctuaries Act was passed to preserve perceived sensitive places in the ocean from wanton destruction, but the act's multiple-use provisions have made it almost impossible to create inviolate havens where the extraction of biotic and abiotic resources is prohibited. In principle, the act was not purely a preservation statute but a legal avenue for the implementation of multiuse sanctuaries. So the result has been that sanctuaries such as Monterey Bay permit commercial and recreational fishing, aquaculture, kelp harvesting, and sand mining to occur within their boundaries.

As sanctuaries came online and managers wrangled with the thorny issue of use, Congress repeatedly failed to address the negative effects of fishing on sanctuaries, content to remain silent on commercial fishing regulations. The task of deciding whether to cover commercial fishing as a regulated or prohibited activity fell to NOAA, which decided not to regulate fishing in the sanctuaries because there was insufficient support for any meaningful restrictions. As a result, resident fish populations within the sanctuaries plummeted. Pacific rockfish, which are structure oriented, favoring rocky outcrops, have been particularly hard hit, experiencing a more than 90 percent decrease in population size over the years. For many managers, the question of at what point multiple uses compromise resource protection merited some serious discussion.

The fundamental flaw of the Sanctuaries Act is its lack of a singular focus on preservation. This omission can be placed at the doorstep of Congress, which has never defined what exactly constitutes a sanctuary and only vaguely identified the act's purpose as protection of special areas of national importance. As more people within the scientific and government communities came to grips with the shortcomings of the Sanctuaries Act, they pondered whether it could be a springboard for the development of future protection schemes.

Marine Protection Areas

In 2000, President Clinton, cognizant that existing protective measures fell short of their intended purpose, issued an executive order calling for federal agencies to study, design, and establish a national system of marine protected areas. The executive order defined a "marine protected area" (MPA) as "any area of the marine environment that has been reserved by Federal, State, territorial, tribal, or local laws or regulations to provide lasting protection for part or all of the natural and cultural resources therein." As detailed in the executive order, the system would include areas of ocean designated to enhance conservation of marine resources, although the level of protection would vary widely.

In practice, there are many types of marine protected areas, ranging from highly restrictive to those that are designed as multiple use. The more restrictive MPAs are marine reserves, known as fully protected or no-take MPAs, which are completely protected from all extractive and destructive activities. Marine reserves, therefore, have explicit prohibitions against fishing and the removal or disturbance of living or nonliving resources.

As stated in Clinton's executive order, MPAs could be established at all levels of government. At the federal level, the national marine sanctuaries are a type of MPA because they allow fishing, extraction, recreation, and other human activities to occur within their boundaries. At the state level, the MPAs are managed by local and/or federal agencies. Regardless of the level of management, the underlying principle of MPAs is that humans cannot manage marine ecosystems but only the activities that occur within them.

California Picks Up the Banner of Conservation

In 1998, a group of citizens from Santa Barbara and Ventura concerned that the resources of the Channel Islands were not receiving adequate protection urged California officials to adopt more-widespread conservation measures. Agreeing that increased human activities were impacting the islands, government agencies arrived at the consensus that one important strategy was to establish marine protected areas. At the same time, the California legislature passed the Marine Life Protection Act, charging the California Fish and Game Commission to implement measures that protect marine habitats and preserve ecosystem integrity. In 2003, based on public input, scientific guidance, and socioeconomic considerations, the commission established a network of marine protected areas within the state waters of the Channel Islands Marine Sanctuary. Ten marine reserves and two marine conservation areas—a less restrictive type of MPA in which a limited take of lobster and pelagic fish was permitted—were designated. In 2006 and 2007, NOAA expanded the MPA network into federal waters. With this expansion, the entire network consisted of eleven marine reserves and two conservation area. The MPAs encompass approximately 21 percent of the Channel Islands Marine Sanctuary, leaving 79 percent open to consumptive recreational and commercial activities regulated by state and federal agencies.

California sheepshead (Joe Belanger/Shutterstock.com)

Have the Channel Islands MPAs, particularly the marine reserve areas, been effective? In 2003, a monitoring plan was initiated with the goal of detecting changes in biology, economic factors, and human activities within and outside the MPAs. Over a five-year period, researchers from a variety of institutions conducted intensive surveys of the marine plants, animals, and habitats of the Channel Islands. While some benefits that accrue with increased protection may not be detected within a five-year period because many species grow slowly and successful reproduction may be infrequent, researchers found a number of positive responses within the biological community: kelp forest abundance and fish biomass, size, and diversity were greater within the boundaries of the marine reserves than in those areas not receiving full protection. The degree of protection depended, in part, on how much time a species spent in the reserve. Tagging studies showed temporal variation among four species: California sheepshead, kelp bass, cabezon, and giant sea bass. Sheepsheads tagged inside a marine reserve at Anacapa Island stayed there for 95 percent of the time, essentially taking up residence. Numerous tagged kelp bass and cabezon stayed in the reserve for extended periods of time, but some exited and did not return. Giant sea bass resided in the reserves only 25 percent of the time they were tracked. It is not surprising, therefore, that the site-faithful California sheepshead showed a large increase in numbers within the reserve.

A secondary effect of the reserves that the researchers noted was a "spillover" of biomass from these sites into neighboring areas. Larger fish are prodigious producers of offspring, so reserves can contribute to fish population replenishment outside their boundaries. In essence, the impact of the reserve may be larger than its actual size. The survey concluded that it was the spatial nature of these reserve areas that confer their benefits when compared with other management measures — focusing on the whole ecosystem rather than attempting to manage one component of the ecosystem.

One nagging question remained: can reserves be beneficial for highly migratory species? Ideally, a protected area would encompass the majority, if not all, of a species's geographic range. In practice, this is not a viable option because many organisms, such as the salmon shark, albacore tuna, and California gray whale, undertake prodigious migrations on an annual basis within and beyond the California Current. Preliminary results showed that even though a highly mobile species might use a protected area for only a limited portion of its life span, some benefits would accrue: reducing the frequency of exposure of the species to threats, such as fishery bycatch, and diminishing the overall cumulative impact of other hazards, such as drilling operations and underwater noise.

Using the success of the Channel Islands MPAs as a springboard, California looked to develop a statewide system of MPAs. This ambitious undertaking was spearheaded by a number of stakeholders, representing a wide range of interests, as opposed to just scientists and regulators. During the design stage, a number of issues had to be addressed about the nature of the MPAs: if too small, the fish swim out; if too big, there is no place for fishing; and if too far apart, there is no cross-fertilization between the MPAs. Though the proposed network of MPAs would stretch along the total length (1,100 miles) of the California coastline, the state was divided up into five distinct regions: north coast (California/Oregon border south to Point Arena), north-central coast (Point Arena to Pigeon Point), central coast (Pigeon Point to Point Conception), south coast (Point Conception to the Mexican border), and San Francisco Bay.

By 2013, with science dictating the establishment of MPAs, California created a new network of 124 conservation areas, which included four categories of MPAs. The most restrictive classification is the state marine reserves (SMRs), which are no-take areas where all extractive activities are prohibited. Within state marine parks (SMPs), recreational fishing is permitted, but commercial take is not allowed. In state marine conservation areas (SMCAs), managers may limit recreational and/or commercial take to protect a specific resource or habitat. State marine recreational management areas (SMRMAs) are designed so that the managing agency may provide, limit, or restrict recreational opportunities to meet other needs that are consistent with the overall theme of conservation and the preservation of resources for future generations. The resulting network designated approximately 10 percent of state waters as SMRs, and about 16 percent of state waters fell within the other categories.

The MPA initiative was not without its share of controversy. Some fishing interests, in particular, vehemently objected to the process, arguing that the closure areas were unnecessary for fisheries already subject to other regulations and would lead to economic hardship. In Southern California, recreational fishermen who fought their lost access suffered a series of defeats in court. After unsuccessfully spending more than $1.3 million on legal challenges, the United Anglers of Southern California had become resigned to the fact that it was fighting a losing battle. Since the expanded MPA network was initiated, the dire predictions of the collapse of the commercial and recreational fishing industries have not materialized. On the other hand, it is too early to determine whether the closures will lead to more biodiversity, bigger fish, and ultimately more fish to catch. And yet, in the Channel Islands, where MPAs have been in place the longest, recreational fishing actually increased in open areas of the islands from 2003 to 2008 as did commercial landings for some species, including squid, urchin, lobster, and crab.

If MPAs and, in particular, SMRs prove in the long run to be effective conservation tools, leading to resource sustainability and ecosystem resilience, can we foresee further expansion? Choosing new reserves will depend, in part, on whether these new additions will complement those already established to form an optimal network. In addition, figurehead species, such as whales and pinnipeds, attract a great deal of attention from environmental and conservation organizations and are often used as a lever to influence proposed legislation. Full protection for these marine megafauna must address the minimum critical habitat necessary for successful breeding and foraging. Historically, protected areas, such as the Gulf of the Farallones National Marine Sanctuary, have been created around the terrestrial habitats used by pinnipeds for hauling out, resting, and breeding, but few examples exist that address the habitat needs of these mammals while they are at sea. The reason for this discrepancy is primarily one of logistics; it is simply easier to encompass large aggregations of individuals in a small protected area. However, the greatest threat to these animals may not be when they are on land; myriad legislation now offers a blanket of protection for most marine mammals from human activities. Greater attention needs to be directed toward their foraging sites, which are often separated spatially from breeding areas, to establish how best to protect their access to food resources.

As we have seen, foraging sites are often associated with specific topographic and oceanographic features. In particular, three types of oceanic "hotspots" have been identified as areas of concern. The first are known as static systems—permanent topographic features, such as submarine canyons, which have an elevated abundance of foraging animals compared with the surrounding areas. Persistent hydrographic features, such as currents and frontal boundaries, are also areas where food is abundant and as such are magnets for predators, some of which migrate great distances to reach these sites. Finally, there are ephemeral habitats, including upwelling regions, eddies, and current filaments, which as the name implies are temporally and spatially variable.

If a particular hotspot occurs within the boundaries of a country's Exclusive Economic Zone, it could, buoyed by the appropriate political will and social commitment, be protected by delimiting a large enough area that would encompass the threatened habitat. On the other hand, habitats occurring outside a country's Exclusive Economic Zone, such as the North Pacific Transition Zone, could be more challenging to protect and manage, requiring international cooperation that historically has been difficult to achieve.

The term "high seas," as adopted by the 1958 Convention on the High Seas, "means all parts of the sea that are not included in the territorial seas or in the internal waters of the State." The convention also codified the rules of international law relating to high seas, which included freedom of fishing. And therein lies the root of the management problem: if the high seas are open to all nations and no state may exert its sovereignty over this area, then its resources are open to the unlimited exploitation by all nations.

A Fly in the Ointment

Regardless of the protective measures and management schemes that are presently in place or might be implemented in the future, their success ultimately hinges on a stable environment, one in which large swings in oceanographic conditions could undermine any hope of achieving long-term conservation goals. Here we need to address the issue of ecosystem resiliency, which can be defined as the capacity of a system to undergo a disturbance without collapsing or metamorphosing into a different state by maintaining its core properties and functions.

In the view of many scientists, one of the greatest threats to ocean resiliency is ocean acidification. Though not receiving the notoriety of an overheated planet or superstorms, ocean acidification may be clearest example of how humans impact the earth's environment.

Acidification occurs when atmospheric carbon dioxide dissolves in seawater, forming carbonic acid, which in turn releases hydrogen ions, reducing the pH (increased acidity). The more carbon dioxide injected into the system, the more acidic the water becomes. While it is highly unlikely that the ocean will ever become an actual acid (fall below a pH of 7), the pH of the ocean surface has already fallen 0.1 unit, representing a 30 percent increase in acidity, due in all likelihood to the increasing amount of carbon dioxide in the atmosphere.

One of the first researchers to sound the alarm about ocean acidification was Richard Feely, a senior scientist at NOAA's Pacific Marine Environmental Laboratory, who during a series of cruises in the 1980s saw the first indication that ocean chemistry was changing. Though he published his findings, his results generated little interest within the broader scientific community—tiny changes in pH in a few isolated areas of a vast ocean were of no big concern. That would change when a series of economic shocks rippled through the Pacific fishing communities.

From 2006 to 2008, shellfish growers up and down the Pacific coast, from California to British Columbia, noted a precipitous drop in the survival of oyster larvae. Something was happening to the larvae at the formative stage of life when they build their shells. No one had an answer for this drastic decline until they tested the water—it was much too acidic. Shellfish bathed in water with a lower pH cannot form shells as easily.

In searching for the "smoking gun" responsible for outbreaks of acidified water, Feely and his colleagues, through extensive water sampling along the Pacific coast, suspected that the corrosive waters they detected at deep depths during their first ocean survey might rise to the surface during seasonal upwelling events. The deep waters become acidified as Ekman transport converges near-surface waters in the subtropical gyre, and that convergence drives downwelling. This corrosive water may remain sequestered far below the surface for decades until it is brought back to the surface. The upwelling that is occurring today along California, Oregon, and Washington coasts is water that was last in contact with the atmosphere during the 1950s and 1960s—a period of far less atmospheric carbon dioxide than at present.

Today, at the peak of the upwelling season, Feely estimates that as much as 30 percent of the water along Pacific coastline has a pH low enough to be corrosive—a result that could have dire economic consequences for the multi-million-dollar shellfish industry. If carbon emissions remain unchecked, then in 50 to 100 years, the upwelled water will be even more corrosive and widespread.

The dramatic changes in ocean chemistry that are now occurring are on a scale not seen in recent history. One has to go back about 59 million years to a geologic period known as the Paleocene-Eocene Thermal Maximum (PETM). During this time, the supercontinent Pangaea was splitting into separate land masses, and it is theorized that huge amounts of carbon were released into the atmosphere and oceans in the form of carbon dioxide and methane. From the analysis of sediment data, geologists James Wright and Morgan Schaller of Rutgers University concluded the carbon was released in the geologic blink of an eye. While they estimated that the globe warmed approximately 9°F in only about thirteen years, the most disruptive effect was likely the exceptional ocean acidification. To many scientists, the rate of acidification during the PETM pales in comparison with what the oceans are experiencing at present. Some estimates place the earlier rate of change at only one-tenth as fast as it is today.

The dramatic changes to Pacific oyster farms may be only the beginning in a series of systemic disruptions to the ocean ecosystem. On a more ominous note is evidence showing that acidification is dissolving the shells of petropods, minuscule free-swimming marine snails that are a stable food source for mackerel, salmon, and herring. When a food source vanishes, the impact ripples up and down the food chain.

What disturbs the scientific community most about the "runaway" ocean acidification is that there is no obvious solution. In the long run, geologic processes, such as rock weathering, will stop the acidification, but long term means tens of thousands of years. What does this mean for the California Current Ecosystem?

A Last-Gasp Solution

There are rumblings emanating from some conservation circles that, under the scenario of a rapidly changing climate, protection must be afforded to those species most in danger. Could aquariums or sea parks become the refuge of threatened species?

Obviously, many organisms do quite well in captivity, but many marine apex species have not fared well. During the 1990s, a team from the University of Miami undertook the delicate task of collecting and keeping alive billfish larvae in a laboratory environment. At first, their efforts were rewarded; they were able to study up close the eighth-of-an-inch larvae. But their success was ephemeral; the tiny billfish lived for only three days in spite of the herculean effort to replicate their natural environment. Even with the lessons learned from this pilot project, researchers still have not been able to grow and rear billfish successfully in captivity. Raising billfish, such as sailfish, spearfish, striped marlin, and blue marlin, in a hatchery setting, like trout, or keeping them captive for long periods of time in an aquarium is most likely not a viable option for solving the plight of climate-stressed billfish.

In 2004, the first juvenile white shark was exhibited at the Monterey Bay Aquarium, prompting Julie Packard, executive director of the aquarium, to trumpet that "it was the most powerful emissary for ocean conservation in our history." Since that groundbreaking exhibit, the aquarium has played host to six more young sharks, some which stayed for months in the aquarium's tanks before being released. While most survived their release, one female shark died a week later after being set free.

The evidence gathered by studying these sharks and other captive great whites appears to be stacked against the use of aquariums as venues for saving these sharks, particularly adults. Simply, they do not fare well in captivity. As open-water species, prone to undertaking long migrations, confinement may lead to "depression," manifesting itself in a number of ways: refusing to eat, head butting the aquarium's glass walls, and exhibiting increased aggression toward their handlers.

Probably no other organism has generated more controversy with regard to its confinement than has the orca. Recent events, such as the killing by an orca named Tilikum of his trainer, Dawn Brancheau, and protests about the tank size of another orca, Lolita, have become hot-button issues.

Whales have been kept in captivity since 1861, when P. T. Barnum put on display a beluga whale that had been captured in the St. Lawrence River in Canada. Barnum was unable to properly care for the animal because of his ignorance of the needs of the whale, and it succumbed within a week of its confinement.

While our understanding of the nature and needs of whales has improved dramatically since Barnum's time, by keeping orcas confined—currently there are dozens living in captivity for public display—both their physical and social environments are altered. In the wild, orcas routinely make dives to hundreds of feet, but due to obvious structural limitations of their pools or pens, orcas do not have this capability, instead spending most of their time at the surface and swimming in circles. When orcas are removed from their habitat, their family bonds are severed, depriving them of social interactions with other members of their species. Confinement fails miserably to replicate an orca's natural existence.

In a rapidly changing environment, are there any other options for government officials, managers, and conservationists committed to ensuring the survivability of marine species as well as whole ecosystems? Unfortunately, at least from my perspective, I don't see a grand solution, and I'm not sure anyone does. Two words come to mind when I ponder the future: hope and adapt. Hope for a better future. Hope for a light at the end of the tunnel. Without hope, would we continue to struggle to protect and preserve our marine resources? Would we continue to sound the alarm that unprecedented changes are occurring in the marine environment? Arguments have been made that organisms will adapt to their new habitats. Some probably will, but others will perish. A look through geologic history tells a grim story of many organisms that roamed the land and swam the early seas but met the fate of extinction.

BIBLIOGRAPHY

Reference Works

Aguilar Soto, N., M. Johnson, P. Madsen, F. Diaz, I. Dominquez, A. Brito, and P. Tyack. "Cheetahs of the Deep Sea: Deep Foraging Sprints in Short-Finned Pilot Whales off Tenerife (Canary Islands)." *Journal of Animal Ecology* 77 (2008): 936–47.

Ainley, D. "Feeding Methods in Seabirds: A Comparison of Polar and Tropical Nesting Communities in the Eastern Pacific Ocean." In *Adaptations within Antarctic Ecosystems*, edited by G. Llano, 669–85. Gautier, Miss.: Gulf Coast Publishing, 1977.

Ainley, D., and R. Boekelheide. *Seabirds of the Farallon Islands*. Palo Alto, CA: Stanford University Press, 1990.

Ainley, D., R. Henderson, H. Huber, R. Boekelheide, S. Allen, and T. McElroy. "Dynamics of White Shark/Pinniped Interactions in the Gulf of Farallones." *Southern California Academy of Sciences Memoirs* 9 (1985): 109–22.

Aksnes, D., and M. Ohman. "Multi-Decadal Shoaling of the Euphotic Zone in the Southern Sector of the California Current System." *Limnology and Oceanography* 54 (2009): 1272–81.

Andrew, R., B. Howe, and J. Mercer. "Ocean Ambient Sound: Comparing the 1960s with the 1990s for a Receiver off the California Coast." *Acoustic Research Letters Online* 3 (2002): 65–70.

Aquinas, T. *Summa Contra Gentiles*. Edited by English Dominican Friars. London: Barnes and Oates, 1924.

Aristotle. *Historia Animalium*. Edited by A. Gotthelf and E. Balme. New York: Cambridge University Press, 2011.

———. *Politics*. In *The Basic Works of Aristotle*, edited by Richard McKeon, 1127–1324. New York: Random House, 1941.

Ayers, J., and M. Lozier. "Physical Controls of the Seasonal Migration of the North Pacific Transition Zone Chlorophyll Front." *Journal of Geophysical Research* 115 (2010). doi:10.1029/2009C00596.

Bakun, A. "Global Climate Change and Intensification of Coastal Ocean Upwelling." *Science* 247 (January 12, 1990): 198–201.

Balance, L., R. Pitman, and P. Fiedler. "Oceanographic Influences on Seabirds and Cetaceans of the Eastern Pacific: A Review." *Progress in Oceanography* 69 (2006): 360–90.

Barkley, R. "The Kuroshio Current." *Science Journal*, March 1970, 54–60.

Baumgartner, T., A. Soutar, and V. Ferreira-Bartrina. "Reconstruction of the History of Pacific Sardine and Northern Anchovy Populations over the Past Two

Millennia from Sediments of the Santa Barbara Basin, California." *CalCOFI Report* 33 (1992): 24–40.

Beale, T. *The Natural History of the Sperm Whale*. London: Littlehampton Book Services Ltd., 1973.

Bedford, D., and F. Hagerman. "The Billfish Fishery Resource of the California Current." *CalCOFI Report* 24 (1983): 70–78.

Bergamin, A. "From Indonesia to California—Protecting Pacific Leatherback Turtles." http://baynature.org/2013/10/13/indonesia-california-protecting -pacific-leatherback-turtles/. January 3, 2014.

Bird, J. "Sperm Whales: The Deep Divers of the Ocean." www.oceanicresearch .org/education/wonders/spermwhales.htm. January 10, 2014.

Blight, L., and A. Burger. "Occurrence of Plastic Particles in Sea-Birds from the Eastern North Pacific." *Marine Pollution Bulletin* 34 (1997): 323–25.

Block, B., et al. "Tracking Apex Marine Predator Movements in a Dynamic Ocean." *Nature* 475 (July 7, 2011): 86–90.

Bodio, S. "The Chumash, Swordfish, and Rock Art." http://stephenbodio.blogspot .com/2005/10/chumash-swordfish-and-rock-art.html. October 30, 2013.

Boisserie, J., F. Lihoreau, and M. Brunet. "The Position of Hippopotamidae within Cetartiodactyla." *Proceedings of the National Academy of Sciences* 102 (2005): 1537–41.

Bonadonna, F., G. Cunningham, P. Jouventon, F. Hesters, and G. Nevitt. "Evidence for Nest-Odour Recognition of Two Species of Diving Petrels." *Journal of Experimental Biology* 206 (2003): 3719–22.

Botsford, L. "Human Activities, Climate Change Affect Marine Populations." *California Agriculture* 51 (1997): 36–44.

Bowen, B., A. Meylan, and J. Avise. "An Odyssey of the Green Sea Turtle." *Proceedings of the National Academy of Sciences* 86 (1989): 573–76.

Boyce, D., M. Lewis, and B. Worm. "Global Phytoplankton Decline over the Past Century." *Nature* 466 (July 26, 2010): 591–96.

Boyd, I., S. Wanless, and C. Camphuysen, eds. *Top Predators in Marine Ecosystems: Their Role in Monitoring and Management*. New York: Cambridge University Press, 2006.

Briggs, K., W. Tyler, D. Lewis, and D. Carlson. *Bird Communities at Sea Off California: Studies in Avian Biology No. 11*. Lawrence, Kans.: Allen Press. 1987.

Brooke, M. "The Food Consumption of the World's Seabirds." *Proceedings of the Royal Society of London: Biology Letters* 271 (2004): S246–48.

Brothers, J., and K. Lohmann. "Evidence for Geomagnetic Imprinting and Magnetic Navigation in the Natal Homing of Sea Turtles." *Current Biology* 25 (February 5, 2015): 1–5.

California Wetfish Producers Association. http://www.californiawetfish.org/. April 6, 2014.

Carr, A., and P. Coleman. "Seafloor Spreading Theory and the Odyssey of the Green Turtle." *Nature* 249 (1974): 128–30.

Carrier, D., S. Deban, and J. Otterstrom. "The Face That Sank the *Essex*: Potential

Function of the Spermaceti Organ in Aggression." *Journal of Experimental Biology* 205 (2002): 1755–63.

Cave, W. "Increased Size and Power for California Tuna Vessels." *Pacific Fisherman* (March 1928): 14.

Champagne, C., D. Crocker, M. Fowler, and D. Houser. "Fasting Physiology of the Pinnipeds: The Challenges of Fasting While Maintaining High Energy Expenditure and Nutrient Delivery for Lactation." In *Comparative Physiology of Fasting, Starvation, and Food Limitation*, edited by M. McCue, 309–33. Berlin: Springer-Verlag, 2012.

Chan, F., J. Barth, J. Lubchenco, A. Krincich, W. Peterson, and B. Menge. "Emergence of Anoxia in the California Current Large Marine Ecosystem." *Science* 319 (February 15, 2008): 920.

Chandler, W., and H. Gillean. "The Makings of the National Marine Sanctuaries Act: A Legislative History and Analysis." http://mcbi.marine-conservation.org/publications/pub_pdfs/The%20Makings%20of%20National%20Marine%20Sanctuaries%20Booklet.pdf. June 6, 2014.

Chase, O. *Shipwreck of the Whale-Ship* Essex. New York: Gilley, 1821.

Checkley, D., and J. Barth. "Patterns and Processes in the California Current System." *Progress in Oceanography* 83 (2009): 49–64.

Clark, F. "The Sardine: International Aspects of Its Life History and Exploitation." *Sixth Pacific Science Congress of the Pacific Science Association* 6 (1939): 35–42.

Cleeland, N. "What Started as a Gamble Is Paying Off for Sportfishing Skippers in San Diego: A Fisherman Can Be Hooked, Too, on a Long Tuna Trip." *Los Angeles Times*, February 11, 1985, 42–43.

Coan, A., M. Vojkovich, and D. Prescott. "The California Harpoon Fishery for Swordfish." *NOAA Technical Report NMFS 142* (1998): 34–49.

Coastal Observation and Seabird Survey Team. "Using Unmanned Aircraft Systems to Monitor Seabirds." http://blogs.uw.edu/coasst/2013/07/08/using-drones-to-monitor-seabirds/. October 11, 2013.

Coleridge, S. *The Rime of the Ancient Mariner*. Mineola, N.Y.: Dover Publications, 1992.

Conniff, R. "An Unlikely Solution: Saving Sea Turtles by Eating Their Eggs." http://www.takepart.com/article/2013/08/12/sea-turtle-eggs-costa-rica-legal-harvest. November 11, 2014.

Connor, R. "Individual Foraging Specialization in Marine Mammals: Culture and Ecology." *Behavior and Brain Science* 24 (2001): 329–30.

Costa, D. "Diving Physiology of Marine Vertebrates." Wiley Online Library. doi:10.1002/9780470015902.a0004230. January 15, 2007.

———. "Energetics." In *Encyclopedia of Life Sciences*, edited by William Perrin, B. Wursig, and J. Thewissen, 383–91. Amsterdam: Elsevier, 2008.

Costa, D., G. Breed, and P. Robinson. "New Insights into Pelagic Migrations: Implications for Ecology and Conservation." *Annual Review of Ecology, Evolution, and Systematics* 43 (2012): 73–96.

Croxall, J. *Seabirds: Feeding Biology and the Role of Marine Ecosystems*. New York: Cambridge University Press, 1987.

Cummings, B. "Pliny's Literate Elephant and the Idea of Animal Language in Renaissance Thought." In *Renaissance Beasts: Of Animals, Humans, and Other Wonderful Creatures*, edited by Erica Fudge, 164–85. Urbana: University of Illinois Press, 2004.

Curry, P., A. Bakun, R. Crawford, A. Jarre, R. Quiñones, L. Shannon, and H. Verheye. "Small Pelagics in Upwelling Systems: Patterns of Interaction and Structural Changes in 'Wasp-Waist' Ecosystems." *ICES Journal of Marine Science* 57 (2000): 603–18.

Darwin, C. *The Voyage of the* Beagle. Oxon: Acheron Press, 2012.

Donohue, M., D. Costa, M. Goebel, and J. Baker. "The Ontogeny of Metabolic Rate and Thermoregulatory Capabilities of Northern Fur Seal, *Callorhinus ursinus*, Pups in Air and Water." *Journal of Experimental Biology* 201 (2000): 1003–16.

Doughty, R. "San Francisco's Nineteenth-Century Egg Basket: The Farallons." *Geography Review* 61 (1971): 554–72.

Dyke, G., and G. Kaiser, eds. *Living Dinosaurs: The Evolutionary History of Modern Birds*. Hoboken, N.J.: Wiley-Blackwell, 2011.

Eckert, S. "Bound for Deep Water." http://www.naturalhistorymag.com/htmlsite /master.html?http://www.naturalhistorymag.com/htmlsite/editors_pick/1992 _03_pick.html. May 4, 2014.

Eckert, S., K. Eckert, P. Ponganis, and G. Kooyman. "Diving and Foraging Behavior of Leatherback Sea Turtles (*Dermochelys coriacea*)." *Canadian Journal of Zoology* 67 (1989): 2834–40.

Ellis, R. *The Great Sperm Whale: A Natural History of the Ocean's Most Magnificent and Mysterious Creature*. Lawrence: University Press of Kansas, 2011.

Erlandson, J., T. Rick, J. Estes, M. Graham, T. Braje, and R. Vellanoweth. "Sea Otters, Shellfish, and Humans: 10,000 Years of Ecological Interaction on San Miguel Island, California." *Proceedings of California Island Symposium* 6 (2005): 58–68.

Estes, J., and D. Duggins. "Sea Otters and Kelp Forests in Alaska: Generality and Variation in a Community Ecological Paradigm." *Ecological Monographs* 65 (1995): 75–100.

Estes, J., and J. Palmisano. "Sea Otters: Their Role in Structuring Nearshore Communities." *Science* 185 (September 20, 1974): 1058–60.

Felando, A., and H. Medina. "The Origins of California's High-Seas Tuna Fleet." https://www.sandiegohistory.org/journal/v58-1/v58-1felando.pdf. May 4, 2014.

Fiedler, P., S. Reilly, R. Hewitt, D. Demer, V. Philbrick, S. Smith, W. Armstrong, D. Croll, B. Tershy, and B. Mate. "Blue Whale Habitat and Prey in the California Channel Islands." *Deep Sea Research Part II: Topical Studies in Oceanography* 45 (1998): 1781–1801.

Finkelstein, M., B. Keitt, D. Croll, B. Tershy, W. Jarman, S. Rodruguez-Pastor, D. Anderson, P. Sievert, and D. Smith. "Albatross Species Demonstrate Regional Differences in North Pacific Marine Contamination." *Ecological Applications* 16 (2006): 678–86.Ford, J., G. Ellis, and M. Graeme. "Selective Foraging by Fish-Eating Killer Whales *Orcinus orca* in British Columbia." *Marine Ecology Progressive Series* 316 (2006): 185–99.

Ford, J., G. Ellis, L. Barrett-Lennard, A. Morton, R. Palm, and K. Balcomb. "Dietary Specialization in Two Sympatric Populations of Killer Whales (*Ornicus orca*) in Coastal British Columbia and Adjacent Waters." *Canadian Journal of Zoology* 76 (1998): 1456–71.

Foster, M., and D. Schiel. "The Ecology of Giant Kelp Forests in California: A Community Profile." *U.S. Fish and Wildlife Service Biological Report* 85 (1985): 1–32.

Francis, D., and G. Hewlett. *Operation Orca: Springer, Luna and the Struggle to Save West Coast Killer Whales.* Madeira Park, British Columbia: Harbour Publishing, 2007.

Francis, R., S. Hare, A. Hollowed, and W. Wooster. "Effects of Interdecadal Climate Variability on the Oceanic Ecosystems of the NE Pacific." *Fisheries Oceanography* 7 (1988): 1–21.

Frazier, I. "Prehistoric and Ancient Historic Interactions between Humans and Marine Turtles." In *The Biology of Sea Turtles*, vol. 2, edited by P. Lutz, J. Musick, and J. Wyneken, 1–38. Boca Raton, Fla.: CRC Press, 2003.

Freely, R., C. Sabine, J. Hernandez-Ayon, D. Ianson, and B. Hales. "Evidence for Upwelling of Corrosive 'Acidified' Water onto the Continental Shelf." *Science* 305 (June 13, 2008): 1490–92.

Fritsches, K., and E. Warrant. "Differences in the Visual Capabilities of Sea Turtles and Blue Water Fishes—Implications for Bycatch Reduction." In *Sea Turtle and Pelagic Fish Sensory Biology: Developing Techniques to Reduce Sea Turtle Bycatch in Longline Fisheries*, edited by Y. Swimmer and R. Brill, 1–7. NOAA Technical Memo, NMFS-PIFSC-7. Washington, D.C.: National Oceanic and Atmospheric Administration, 2006.

Furness, R., and K. Camphuysen. "Seabirds as Monitors of the Marine Environment." *ICES Journal of Marine Science* 54 (1997): 726–37.

Furness, R., and P. Monaghan. *Seabird Ecology.* Glasgow: Blackie, 1987.

Gangopadhyay, A., P. Lermusiaux, L. Rosenfeld, A. Robinson, L. Caldo, H. Kim, W. Leslie, and P. Haley. "The California Current System: A Multiscale Overview and the Development of a Feature-Oriented Regional Modeling System (FORMS)." *Dynamics of Atmospheres and Oceans* 52 (2011): 131–69.

Garcia-Parraga, D., J. Crespo-Picazo, Y. de Quiros, V. Cervera, L. Marti-Bonmati, J. Diaz-Delagardo, M. Arbelo, M. Moore, P. Jepson, and A. Fernandez. "Decompression Sickness ('the Bends') in Sea Turtles." *Diseases of Aquatic Organisms* 111 (2014): 191–205.

Gazda, S., R. Connor, R. Edgar, and F. Cox. "A Division of Labour in Group-Hunting Bottlenose Dolphins (*Tursiops truncatus*) off Cedar Key, Florida." *Proceedings of the Royal Society B* 272 (2005): 135–40.

Gerard, V. "Growth and Utilization of Internal Nitrogen Reservoirs by the Giant Kelp *Macrocystis pyrifera* in a Low-Nitrogen Environment." *Marine Biology* 66 (1982): 27–35.

Gerlinsky, C., D. Rosen, and A. Trites. "High Diving Metabolism Results in Short Aerobic Dive Limit for Steller Sea Lions (*Eumetopias jubatus*)." *Journal of Comparative Physiology B* 183 (2013): 699–708.

Gingerich, P., S. Raza, M. Arif, M. Anwar, and X. Zhou. "New Whale from the Eocene of Pakistan and the Origin of Cetacean Swimming." *Nature* 368 (1994): 844–47.

Goebel, N., C. Edwards, J. Zehr, M. Follows, and S. Morgan. "Modeled Phytoplankton Diversity and Productivity in the California Current Ecosystem." *Ecological Modelling* 264 (2013): 37–47.

Goldbogen, J. "The Ultimate Mouthful: Lunge Feeding in Rorqual Whales." *American Scientist* 98 (March–April 2010): 124–30.

Graham, M., B. Kinlan, and R. Grosberg. "Post-Glacial Redistribution and Shifts in Productivity of Giant Kelp Forests." *Proceedings of the Royal Society B* 277 (2010): 399–407.

Grey, Z. *Tales of Fishes*. New York: Harper and Brothers, 1919.

Griebel, U., and L. Peichl. "Colour Vision in Aquatic Mammals—Facts and Open Questions." *Aquatic Mammals* 29 (2003): 18–30.

Handler, N. "Biogeochemical and Ecological Provinces within the California Current System." http://www.mbari.org/education/internship/02interns /02papers/nicholas.pdf. February 22, 2014.

Haney, C., and A. Stone. "Seabird Foraging Tactics and Water Clarity: Are Plunge Divers Really in the Clear." *Marine Ecology* 49 (1988): 1–9.

Harting, J. *Ornithology of Shakespeare*. New York: Hyperion Books, 1978.

Hazen, E., S. Jorgensen, R. Rykaczewski, S. Bogard, D. Foley, I. Jonsen, S. Shaffer, J. Dunne, D. Costa, L. Crowder, and B. Block. "Predicted Habitat Shifts of Pacific Top Predators in a Changing Climate." *Nature Climate Change* 3 (2013): 234–38.

Hemila, S., S. Nummela, A. Berta, and T. Reuter. "High-Frequency in Phocid and Otariid Pinnipeds: An Interpretation Based on Inertial and Cochlear Constraints." *Journal of the Acoustical Society of America* 120 (2006): 3463–66.

Hickey, B. "The California Current System—Hypotheses and Fact." *Progress in Oceanography* 8 (1979): 191–279.

———. "Circulation over the Santa Monica–San Pedro Basin and Shelf." *Progress in Oceanography* 30 (1992): 37–115.

Hodder, J., and M. Graybill. "Reproduction and Survival of Seabirds in Oregon during the 1982–1983 El Niño." *Condor* 87 (1985): 535–41.

Hooker, S. "Marine Reserves and Higher Predators." In *Top Predators in Marine Ecosystems: Their Role in Monitoring and Management*, edited by I. Boyd, S. Wanless, and C. Camphuysen, 347–59. New York: Cambridge University Press, 2006.

Hooker, S., and L. Gerber. "Marine Reserves as a Tool for Ecosystem-Based Management: The Potential Importance of Megafauna." *BioScience* 54 (2004): 27–39.

Hooker, S., et al. "Deadly Diving? Physiological and Behavioural Management of Decompression Stress in Diving Mammals." *Proceedings of the Royal Society B* 279 (2012): 1041–60.

Howell, E., D. Kobayashi, D. Parker, G. Balazs, and J. Polovina. "Turtle Watch: A Tool to Aid in the Bycatch Reduction of Loggerhead Turtles *Caretta caretta* in the Hawaii-Based Pelagic Longline Fishery." *Endangered Species Research* 5 (2008): 267–78.

Hyrenbach, K., K. Forney, and P. Dayton. "Marine Protected Areas and Ocean Basin Management." *Aquatic Conservation: Marine and Freshwater Ecosystems* 10 (2000): 437–58.

Jobling, J. *A Dictionary of Scientific Bird Names*. New York: Oxford University Press, 1991.

Johansson, L., and B. Aldrin. "Kinematics of Diving Atlantic Puffins (*Fratercula arctica*): Evidence for an Active Upstroke." *Journal of Experimental Biology* 205 (2002): 371–78.

Jones, J., and I. Jones. "Western Boundary Current in the Pacific: The Development of Our Oceanographic Knowledge." In *Oceanography History: The Pacific and Beyond*, edited by K. Benson and P. Rehbock, 86–95. Seattle: University of Washington Press, 2002.

Joy, R. "How the Sea Otter Hunt Began." http://www.fortrossstatepark.org/seaotter .htm. December 4, 2014.

Kastak, D., and R. Schusterman. "Low-Frequency Amphibious Hearing in Pinnipeds: Methods, Measurements, Noise, and Ecology." *Journal Acoustic Society of America* 103 (1998): 2216–28.

Kennedy, C. "An Upwelling Crisis: Ocean Acidification." http://www.climate.gov /news-features/features/upwelling-crisis-ocean-acidification. June 3, 2014.

Kimball, L. "The Protection of the Marine Environment: A Key Policy Element." In *Ocean Management in Global Change: Proceedings of the Conference on Ocean*, edited by P. Fabbri, 329–44. Boca Raton, Fla.: CRC Press, 1990.

Klimley, A. "The Predatory Behavior of the White Shark." *American Scientist* 22 (March–April 1994): 122–33.

Kooyman, G., M. Castellini, and R. Davis. "Physiology of Diving in Marine Mammals." *Annual Review of Physiology* 43 (1981): 343–56.

Koslow, J., R. Goericke, A. Lara-Lopez, and W. Watson. "Impact of Declining Intermediate-Water Oxygen on Deepwater Fishes in the California Current." *Marine Ecology Progress Series* 436 (2011): 207–18.

Kosro, P., A. Huyer, S. Ramp, R. Smith, F. Chavez, T. Cowles, M. Abbott, P. Strub, R. Barber, P. Jessen, and L. Small. "The Structure of the Transition Zone between Coastal Waters and the Open Ocean off Northern California, Winter and Spring 1987." *Journal of Geophysical Research* 96 (1991): 14, 707–30.

Kovner, G. "Underwater Mountain of Cordell Bank National Marine Sanctuary Is Haven for Sea Life." http://www.pressdemocrat.com/news/2213203–181 /underwater-mountain-of-cordell-bank. May 8, 2014.

Laurs, R., and R. Lynn. "North Pacific Albacore Ecology and Oceanography." In *NOAA Tech. Report NMFS 105*, edited by J. Wetherall, 69–79. Washington, D.C.: National Oceanographic Atmospheric Administration, December 1991.

Levinton, J. *Marine Biology: Function, Biodiversity, Ecology*. New York: Oxford University Press, 1995.

Lewison, R., S. Freeman, and L. Crowder. "Quantifying the Effects of Fisheries on Threatened Species: The Impact of Pelagic Longlines on Loggerhead and Leatherback Sea Turtles." *Ecology Letters* 7 (2004): 221–31.

Lichatowich, J. *Salmon without Rivers: A History of the Pacific Salmon Crisis.* Washington, D.C.: Island Press, 1999.

Lihoreau, F., J. Boisserie, F. Manthi, and S. Ducrocq. "Hippos Stem from the Longest Sequence of Terrestrial Cetartiodactyl Evolution in Africa." *Nature Communications* 6 (2015). doi:10.1038/ncomms7264.

Liwanag, H., A. Berta, D. Costa, S. Budge, and T. Williams. "Morphological and Thermal Properties of Mammalian Insulation: The Evolutionary Transition to Blubber in Pinnipeds." *Biological Journal of the Linnean Society* 107 (2012): 774–87.

Lluch-Belda, D., R. Schwartzlose, R. Serra, R. Parrish, T. Kawasaki, D. Hedgecock, and R. Crawford. "Sardine and Anchovy Regime Fluctuations of Abundance in Four Regions of the World Oceans: A Workshop Report." *Fisheries Oceanography* 1 (1992): 339–47.

Lohmann, K., and C. Lohmann. "Orientation to Ocean Waves by Green Turtle Hatchlings." *Journal of Experimental Biology* 177 (1992): 1–13.

Lohmann, K., N. Putnam, and C. Lohmann. "Geomagnetic Imprinting: A Unifying Hypothesis of Long-Distance Migration in Salmon and Sea Turtles." *Proceedings of the National Academy of Sciences* 105 (2008): 19096–101.

Lubchenco, J., S. Palumbi, S. Gaines, and S. Andelman. "Plugging a Hole in the Ocean: The Emerging Science of Marine Reserve." *Ecological Applications* 13 (2003): 3–7.

Lynam, C., M. Gibbons, B. Axelsen, C. Sparks, J. Coetzee, B. Heywood, and A. Bierley. "Jellyfish Overtake Fish in a Heavily Fished Ecosystems." *Current Biology* 16 (2006): 8492–93.

Lynn, R., and J. Simpson. "The California Current System: The Seasonal Variability of Its Physical Characteristics." *Journal of Geophysical Research* 92 (September 15, 1987): 12947–66.

Malmquist, D. "The Great Ocean Migration." *Currents*, Spring 1997, 14–18.

Mansfield, K., J. Wyneken, W. Porter, and J. Luo. "First Satellite Tracks of Neonate Sea Turtles Redefine the 'Lost Years' Oceanic Niche." *Proceedings of the Royal Society B.* http://rspb.royalsocietypublishing.org/content/281/1781/20133039.full.pdf+html. March 1, 2014.

Mantua, N. "Pacific-Decadal Oscillation." In *Encyclopedia of Global Environmental Change*, edited by M. MacCracken and J. Perry, 592–94. New York: John Wiley and Sons, 2002.

Mantua, N., S. Hare, Y. Zhang, J. Wallace, and R. Francis. "A Pacific Interdecadal Climate Oscillation with Impacts on Salmon Production." *Bulletin American Meteorological Society* 78 (1997): 1069–79.

Maury, M. *Physical Geography of the Sea, and its Meteorology.* London: Forgotten Books, 2012.

Mayo, C., and M. Marx. "Surface Foraging Behaviour of the North Atlantic Right Whale, *Eubalaena glacialis*, and Associated Zooplankton Characteristics." *Canadian Journal of Zoology* 68 (1990): 2214–20.

Mazouchova, N., N. Gravish, A. Savu, and D. Goldman. "Utilization of Granular Solidification during Terrestrial Locomotion of Hatchling Sea Turtles." *Biology Letters* 6 (June 23, 2010): 396–401.

McKown, M. "Acoustic Communication in Colonial Seabirds: Individual, Sexual, and Species-Specific Variation in Acoustic Signal of *Pterodroma* Petrels." Ph.D. diss., University of North Carolina, 2008.

Michigan State University. "Seabird Bones Reveal Changes in Open-Ocean Food Chain." http://msutoday.msu.edu/news/2013/seabird-bones-reveal-changes-in -open-ocean-food-chain/. May 13, 2014.

Milner, A., and S. Walsh. "Avian Brain Evolution: New Data from Palogene Birds (Lower Eocene) from England." *Zoological Journal of Linnean Society* 155 (2009): 198–219.

Milton, J. *Paradise Lost*. New York: Barnes and Noble Classics, 2004.

Monterey Bay Aquarium Foundation. "Giant Kelp." http://www.montereybay aquarium.org/animal-guide/plants-and-algae/giant-kelp. August 12, 2015.

Moore, J., and J. Barlow. "Declining Abundance of Beaked Whales (Family Ziphiidae) in the California Current Large Marine Ecosystem." *PLoS ONE*. doi:10.1371/journal.pone.0052770. January 16, 2013.

Moore, M. "California—Where Have All the Albacore Gone." http://www .fishermensnews.com/story/2013/12/01/features/california-where-have-all -the-albacore-gone/227.html. June 6, 2014.

Naess, C. "How Much Food Do Seabirds Need?" http://sciencenordic.com/how -much-food-do-seabirds-need. November 12, 2013.

National Oceanic and Atmospheric Administration. "The Amazing Seabirds of Cordell Bank National Marine Sanctuary." http://cordellbank.noaa.gov/library /seabirds_guide.pdf. November 22, 2013.

Nevitt, G., M. Losekoot, and H. Weimerskirch. "Evidence for Olfactory Search in Wandering Albatross, *Diomedea exulans*." *Proceedings of the National Academy of Sciences* 105 (2008): 4576–81.

Nur, N., J. Jahncke, M. Herzog, J. Howar, K. Hyrenbach, J. Zamon, D. Ainley, J. Wiens, K. Morgan, L. Balance, and D. Stralberg. "Where the Wild Things Are: Predicting Hotspots of Seabird Aggregation." *Ecological Applications* 21 (2011): 2241–57.

Oedekoven, C., D. Ainley, and L. Spear. "Variable Response of Seabirds to Change in Marine Climate: California Current, 1985–1994." *Marine Ecology Progress Series* 212 (2001): 265–81.

Ogden, A. *The California Sea Otter Trade, 1784–1848*. Berkeley: University of California Press, 1975.

Owen, R. "Eddies of the California Current System: Physical and Ecological Characteristics." *2nd California Islands Multi-disciplinary Symposium* (1980): 237–62.

Pabst, D., S. Rommel, and B. McLellan. "The Functional Morphology of Marine Mammals." In *Biology of Marine Mammals*, edited by J. Reynolds and S. Rommel, 15–72. Washington, D.C.: Smithsonian Books, 1999.

Palmer, R. "California Tuna Canning Industry." *Pacific Fisherman Yearbook* (January 1915): 76–77.

Parker, J. *A Great Sign of Land: Columbus and Sea-Birds: Ornithology and Navigation in 1492*. Minneapolis, Minn.: Cleora Press, 1992.

Parsons, E. *An Introduction to Marine Mammal Biology and Conservation*. Burlington, Mass.: Jones and Bartlett Learning, 2012.

Pauly, D., V. Christensen, J. Dalsgaard, R. Frosse, and F. Torres. "Fishing Down Marine Food Webs." *Science* 279 (1998): 860–63.

Pearcy, W. "Biology of the Transition Zone." In *NOAA Technical Report NMFS 105*, edited by J. Wetherall, 39–55. Washington, D.C.: National Oceanographic and Atmospheric Administration, December 1991.

———. "Marine Nekton Off Oregon and the 1997–98 El Nino." *Progress in Oceanography* 54 (2002): 399–403.

Perry, P. "The History of the Makah Whale Hunt." http://www.alamut.com/subj/the_other/misc/makahWhaling.html. May 23, 2014.

PEW Environment Group. *The State of Science: Forage Fish in the California Current*. Washington, D.C.: PEW Environment Group, January 2013.

Piatt, J., and R. Ford. "How Many Seabirds Were Killed by the *Exxon Valdez* Oil Spill." *American Fisheries Society Symposium* 18 (1996): 712–19.

Pitman, R. "Seabird Associations with Marine Turtles in the Eastern Pacific Ocean." *Colonial Waterbirds* 16 (1993): 194–201.

Platt, J., W. Sydeman, and F. Wiese. "Introduction: A Modern Role for Seabirds as Indicators." *Marine Ecology Progress Series* 352 (2007): 199–204.

Pliny the Elder. *Natural History*. New York: Oxford University Press, 2004.

Polovina, J., E. Howell, D. Kobayashi, and M. Seki. "The Transition Zone Chlorophyll Front Updated: Advances from a Decade of Research." *Progress in Oceanography* (2015). doi:10.1016/j.pocean.2015.01.00g. January 23, 2015.

Polybius. *Histories*. Translated by E. Shuckburgh. New York: Macmillan, 1889.

Pourade, R. *The Rising Tide: The History of San Diego*. San Diego, Calif.: Union-Tribune Publishing, 1967.

Prince, E., and C. Goodyear. "Hypoxia-Based Habitat Compression of Tropical Pelagic Fishes." *Fisheries Oceanography* 15 (2006): 451–64.

Pyenson, N., J. Goldbogen, A. Vogl, G. Szathmary, R. Drake, and R. Shadwick. "Discovery of a Sensory Organ That Coordinates Lunge Feeding in Rorqual Whales." *Nature* 485 (May 24, 2012): 498–501.

Radovich, J. "The Collapse of the California Sardine Fishery—What Have We Learned." In *Resource Management and Environmental Uncertainty: Lessons from Coastal Upwelling Fisheries*, edited by M. H. Glantz and J. D. Thompson, 107–36. New York: John Wiley and Sons, 1981.

Ramprashad, F., S. Corey, and K. Ronald. "Anatomy of the Seal's Ear." In *Functional Anatomy of Marine Mammals*, edited by R. Harrison, 263–306. London: Academic Press, 1972.

Rathbun, G., B. Hatfield, and T. Murphy. "Status of Translocated Sea Otters at San Nicholas Island, California." *Southwestern Naturalist* 45 (2000): 322–75.

Regular, P., A. Hedd, and W. Montevecchi. "Fishing in the Dark: A Pursuit-Diving

Seabird Modifies Foraging Behaviour in Response to Nocturnal Light Levels."
PLoS ONE 6 (2011). doi:10.1371/journal.pone.0026763.

Reichmuth, C., M. Holt, J. Muslow, J. Sills, and B. Southall. "Comparative Assessment of Amphibious Hearing in Pinnipeds." *Journal of Comparative Physiology A* 199 (2013):491–507.

Reinstedt, R. *Where Have All the Sardines Gone?* Carmel, Calif.: Ghost Town Publishers, 1978.

Repenning, C. "Warm-Blooded Life in Cold Ocean Currents." *Oceans* 13 (1980): 18–24.

Richardson, W. "Fishermen of San Diego: The Italians." *Journal of San Diego History* 27 (1981): 84–94.

Ridgeway, S., and A. Hanson. "Sperm Whales and Killer Whales with the Largest Brains of All Toothed Whales Show Extreme Differences in Cerebellum." *Brain, Biology and Evolution.* doi:10.1159/000360519, 2014.

Roberts, S., and K. Brink. "Managing Marine Resources: Sustainability." *Environment* 52 (July/August 2010): 44–52.

Roden, G. "Subarctic-Subtropical Transition Zone of the North Pacific: Large-Scale Aspects and Mesoscale Structure." In *NOAA Technical Report NMFS 105,* edited by J. Wetherall, 1–38. Washington, D.C.: National Oceanographic and Atmospheric Administration, December 1991.

Roemmich, D., and J. McGowan. "Climatic Warming and the Decline of Zooplankton in the California Current." *Science* 267 (March 3, 1995): 1324–26.

Rogers, P. "Unusual Warm Ocean Conditions off California, West Coast Bringing Odd Species." http://www.mercurynews.com/science/ci_26851300/unusual -warm-ocean-conditions-off-california-west-coast. November 2, 2013.

Rosefield, H. "Bird of the Month: The Albatross." http://the-toast.net/2013/09/26 /bird-month-albatross/. March 22, 2014.

Rosen, D., and A. Trites. "Changes in Metabolism in Response to Fasting and Food Restriction in the Steller Sea Lion (*Eumetopias jubatus*)." *Comparative Biochemistry and Physiology Part B* 132 (2002): 389–99.

Rothschild, B., and V. Naples. "Decompression Syndrome and Diving Behavior in *Odontocheyls*, the First Turtle." *Acta Palaeontoligica Polonica.* http://dx.doi.org /10.4202/app. 2012.0133. March 11, 2014.

Safina, C. *Eye of the Albatross: Visions of Hope and Survival.* New York: Henry Holt, 2002.

Scammon, C. *The Marine Mammals of the Northwestern Coast of North America.* Berkeley: Heyday, 2007.

Scavia, D., et al. "Climatic Change Impacts on U.S. Coastal and Marine Ecosystems." *Estuaries* 25 (2002): 149–64.

Schafer, D. "Eating Turtles in Ancient China." *Journal of American Oriental Society* 82 (1962): 73–74.

Schneider, B. "Turtle Hitchhikers." http://ysm.research.yale.edu//article.jsp ?articleID=383. January 16, 2014.

Schofield, W. "Sardine Fishing Methods at Monterey, California." *Division of Fish and Game of California Fish Bulletin No. 19* (1929): 1–68.

Schorr, G., E. Falcone, D. Moretti, and R. Andrews. "First Long-Term Behavioral Records from Cuvier's Beaked Whale (*Ziphius cavirostris*) Reveal Record-Breaking Dive." *PLoS ONE* 9 (2014). doi:10.1371/journal.pone.0092633.

Schreiber, E., and J. Burger. *Biology of Marine Birds*. Boca Raton, Fla.: CRC Press, 2001.

Schusterman, R., D. Kastak, D. Levenson, C. Reichmuth, and B. Southall. "Why Pinnipeds Don't Echolocate." *Journal Acoustical Society of America* 107 (2000): 2256–74.

Schwing, F., D. Palacios, and S. Bogard. "El Niño Impacts on the California Current Ecosystem." *U.S. CLIVAR Newsletter* 3 (2005): 5–8.

Sea Turtle Conservancy. "Information about Sea Turtles: Threats from Harvest for Consumption." http://www.conserveturtles.org/seaturtleinformation.php?page=harvest. October 29, 2013.

Shaffer, S., Y. Tremblay, H. Weimerskirch, D. Scott, D. Thompson, P. Sagar, H. Moller, G. Taylor, D. Foley, B. Block, and D. Costa. "Migratory Shearwaters Integrate Oceanic Resources across the Pacific in an Endless Summer." *Proceedings of the National Academy of Sciences* 103 (2006): 12799–802.

Shillinger, G., D. Palacios, H. Bailey, S. Bogard, A. Swithenbank, P. Gaspar, B. Wallace, J. Spotila, F. Paladino, R. Pierda, S. Eckert, and B. Block. "Persistent Leatherback Turtle Migrations Present Opportunities for Conservation." *PLoS Biology* 6 (2008). doi:10.1371/journal.pbio.0060171.

Shore, T. "It's Time to Stop Drift Gillnet Fishing off the California Coast." http://www.earthisland.org/journal/index.php/elist/eListRead/its_time_to_stop_drift_gillnet_fishing_off_the_california_coast/. July 7, 2014.

Skogsberg, T. "Hydrography of Monterey Bay, California. Thermal Conditions, 1929–1933." *Transactions of the American Philosophical Society* 29 (1936): 1–152.

———. *Preliminary Investigation of the Purse Seine Industry of Southern California*. Sacramento, Calif.: California State Print Office, 1925.

Skogsberg, T., and P. Phelps. "Hydrography of Monterey Bay, California: Thermal Conditions, Part II, 1934–1937." *Proceedings of the American Philosophical Society* 90 (1946): 350–86.

Smith, J. "Anatomy of a Tuna Clipper." http://www.sandiegoreader.com/news/2013/jan/09/unforgettable-anatomy-tuna-clipper/#. July 9, 2014.

Smith, R., P. Dustan, D. Au, S. Baker, and E. Dunlap. "Distribution of Cetaceans and Sea-Surface Chlorophyll Concentrations in the California Current." *Marine Biology* 91 (1986): 385–402.

Snyderman, M. "By Air and Sea: The Amphibious Lives of Seabirds." http://www.dtmag.com/Stories/Marine%20Life/07–05-whats_that.htm. November 4, 2014.

Steele, J., and E. Henderson. "Coupling between Physical and Biological Scales." *Philosophical Transactions of the Royal Society of London* 343 (1994): 5–9.

Steinbeck, J. *Cannery Row*. New York: Viking Press, 1945.

———. *The Log from the Sea of Cortez*. New York: Viking Press, 1941.

Tasker, M. C. Camphuysen, J. Cooper, S. Garthe, W. Montevecchi, and S. Blaber. "The Impacts of Fishing on Marine Birds." *ICES Journal of Marine Science* 57 (2000): 531–47.

Tierney, L. "Detailed Discussion of Laws concerning Orcas in Captivity." https://www.animallaw.info/article/detailed-discussion-laws-concerning-orcas-captivity. June 13, 2014.

Traugott, J., A. Nesterova, and G. Sachs. "The Nearly Effortless Flight of the Albatross." http://spectrum.ieee.org/aerospace/robotic-exploration/the-nearly-effortless-flight-of-the-albatross. September 3, 2013.

Ueber, E., and A. MacCall. "The Rise and Fall of the California Sardine Empire." In *Climate Variability, Climate Change, and Fisheries*, edited by M. H. Glantz, 31–48. New York: Cambridge University Press, 1992.

Ulanski, S. *The Billfish Story: Swordfish, Sailfish, Marlin, and Other Gladiators of the Sea*. Athens: University of Georgia Press, 2013.

Velarede, E., E. Ezcurra, and D. Anderson. "Seabird Diets Early Warnings of Sardine Fishery Declines in the Gulf of California." *Scientific Reports* 3 (2013). doi:10.1038/ srep01332.

Velt, R., P. Pyle, and J. McGowan. "Ocean Warming and Long-Term Change in Pelagic Bird Abundance within the California Current System." *Marine Ecology Progress Series* 139 (1996): 11–18.

Vickery, J., and M. Brooke. "The Kleptoparasitic Interactions between Great Frigatebirds and Masked Boobies on Henderson Island, South Pacific." *Condor* 96 (1994): 331–40.

Vojkovich, M. "The California Fishery for Market Squid." *CalCOFI Report* 39 (1998): 55–60.

Wang, J., J. Barkan, S. Fisler, C. Godinez-Reyes, and Y. Swimmer. "Developing Ultraviolet Illumination of Gillnets as a Method to Reduce Sea Turtle Bycatch." *Biology Letters* 9 (2013): 75–80.

Watanabe, Y., A. Takahashi, K. Sato, M. Viviant, and C. Bost. "Poor Flight Performance in Deep-Diving Cormorants." *Journal of Experimental Biology* 214 (2011): 412–21.

Watson, K. "'Dolphin Safe' Labels on Canned Tuna Are a Fraud." http://www.forbes.com/sites/realspin/2015/04/29/dolphin-safe-labels-on-canned-tuna-are-a-fraud/. May 17, 2015.

Webb, R. *On the Northwest: Commercial Whaling in the Pacific Northwest, 1790–1967*. Vancouver: University of British Columbia Press, 1980.

Whitehead, H., J. Gordon, E. Matthews, and K. Richard. "Obtaining Skin Samples from Living Sperm Whales." *Marine Mammal Science* 6 (1990): 316–26.

Wiley, D., C. Ware, A. Bocconcelli, D. Cholewiak, A. Friedlander, M. Thompson, and M. Weinrich. "Underwater Components of Humpback Whale Bubble-Netting Feeding Behaviour." *Behaviour* 148 (2011): 575–602.

Williams, T., J. Estes, D. Doak, and A. Springer. "Killer Appetites: Assessing the Role of Predators in Ecological Communities." *Ecology* 85 (2004): 3733–84.

Wilson, E., K. Miller, D. Allison, and M. Magliocca. "Why Healthy Oceans Need Sea Turtles: The Importance of Sea Turtles to Marine Ecosystems." http://oceana.org/sites/default/files/reports/Why_Healthy_Oceans_Need_Sea_Turtles.pdf. December 12, 2014.

Wilson, R. "Fisheries at Risk as Oceans Acidify." *Washington Post*, July 31, 2014, A3.

Witherington, B. *Sea Turtles: An Extraordinary Natural History of Some Uncommon Turtles*. New York: Voyageur Press, 2006.

Witzell, W. "Selective Predation on Large Chelonid Sea Turtles by Tiger Sharks (*Galeocerdo cuvier*)." *Japanese Journal of Herpetology* 12 (1987): 22–29.

Wright, J., and M. Schaller. "Evidence for a Rapid Release of Carbon at the Paleocene-Eocene Thermal Maximum." *Proceedings of the National Academy of Sciences* 110 (2013): 15908–13.

Wyneken, J. "Sea Turtle Locomotion: Mechanisms, Behavior, and Energetics." In *The Biology of Sea Turtles*, edited by P. Lutz and J. Musick, 165–98. Boca Raton: Fla.: CRC Press, 1996.

Zeidberg, L., and B. Robison. "Invasive Range Expansion by the Humboldt Squid, *Dosidicus gigas*, in the Eastern North Pacific." *Proceedings of the National Academy of Sciences* 104 (2007): 12948–50.

Zeidberg, L., W. Hammer, N. Nezlin, and A. Henry. "The Fishery for California Market Squid (*Loligo opalescens*) (Cephalopoda: Myopsida), from 1981 through 2003." *Fishery Bulletin* 104 (2008): 46–59.

Zieralski, E. "Anglers Will Pay for Planned Research into Mexico's Revillagigedos, Where They Will Release All Yellowfin and Wahoo." *San Diego Union-Tribune*, October 8, 2005, B7–8.

Zubryd, S. "The Salmon Snatchers." http://sciencenotes.ucsc.edu/2011/pages /salmon/salmon.html. June 30, 2014.

Zydelis, R., C. Small, and G. French. "The Incidental Catch of Seabirds in Gillnet Fisheries: A Global View." *Biological Conservation* 162 (2013): 76–88.

Guides

Behnke, R. *Trout and Salmon of North America*. New York: Free Press, 2002.

Carwardine, M. *Whales, Dolphins and Porpoises*. London: DK Adult, 2002.

Compagno, L., M. Dando, and S. Fowler. *Sharks of the World*. Princeton, N.J.: Princeton University Press, 2005.

Eschmeyer, W., E. Herald, and H. Hammann. *A Field Guide to Pacific Coast Fishes of North America*. Boston: Houghton Mifflin Company, 1983.

Ford, J., G. Ellis, and K. Balcomb. *Killer Whales*. Vancouver, British Columbia: UBC Press, 2000.

Howell, G., and B. Sullivan. *Offshore Sea Life ID Guide: West Coast*. Princeton Field Guides. Princeton, N.J.: Princeton University Press, 2015.

Kozloff, E. *Marine Invertebrates of the Pacific Northwest*. Seattle: University of Washington Press, 1996.

Pepperell, J. *Fishes of the Open Ocean: A Natural History and Illustrated Guide*. Chicago: University of Chicago Press, 2010.

Perrine, D. *Sea Turtles of the World*. Minneapolis, Minn.: Voyageur Press, 2003.

Richards, A. *Seabirds of the Northern Hemisphere*. New York: Gallery Books, 1990.

Riedman, M. *The Pinnipeds: Seals, Sea Lions, and Walruses*. Berkeley: University of California Press, 1990.

Schirihai, H., and B. Jarrett. *Whales, Dolphins, and Other Marine Mammals.*
    Princeton, N.J.: Princeton University Press 2006.
Schultz, K. *Field Guide to Saltwater Fish.* New York: John Wiley and Sons, 2004.
Spotila, J. *Sea Turtles: A Complete Guide to Their Biology, Behavior, and
    Conservation.* Baltimore: Johns Hopkins University Press, 2004.
Stewart, B., P. Clapham, J. Powell, and R. Reeves. *National Audubon Society Guide
    to Marine Mammals of the World.* New York: Knopf, 2002.

INDEX